BEYOND HIGH TECH SURVIVAL

turning government policy into
international profits

by cliff jernigan

TEN WAYS THAT YOU CAN INCREASE YOUR INTERNATIONAL PROFITS

> **Become familiar with U.S. trade laws,** especially those which affect your industry, and learn how to effectively manage them. This knowledge could double your sales to overseas markets. In the past 10 years, the U.S. semiconductor industry has been able to double its sales to Japan because of U.S. trade laws designed to help open foreign markets.

> **Be aware of U.S. anti-dumping laws** and how they affect your industry. Become proactive in conjunction with other similar businesses to protect yourselves. In the mid-1980s, many U.S. companies were forced out of the semiconductor memory business because of foreign dumping costing these companies about 2 billion dollars in sales. Don't let this happen to you.

> **Learn why export control regulations** on certain products may severely hamper your ability to sell overseas. Research those relevant to your industry, and again, become proactive. Currently, the U.S. software industry is feeling the impact of these regulations and is waging a tough battle with the U.S. Congress and Administration to save billions of dollars of overseas sales.

> **Fend off frivolous security lawsuits** that could put a major dent in your bottom line. Find out how companies are lobbying the government to stop this abusive practice.

> **Protect your company's patents,** copyrights and trade secrets. Theft of these items could spell disaster. Find out what tactics companies use to stop the piracy of such property.

> **Increase your R&D spending and enjoy a healthy tax break.** The U.S. tax code provides generous tax incentives for companies doing research and development (a dollar for dollar credit for a portion of the increase to be taken off your total tax bill). However, knowing how to make the law work for you may be a little tricky. Find out how industry lobbied Congress to make the law work for them.

> **Avoid sexual harassment** and wrongful termination lawsuits. They not only damage your company's image, they can also affect your bottom line. See why companies usually settle and what steps they take to avoid future lawsuits.

> **Learn why U.S. manufacturing subsidiaries overseas** generally pay little or no current U.S. taxes. There have been proposals in the U.S. Congress to abolish this tax break. Learn what you can do to maintain the status quo.

> **Export, Export, Export!** U.S. law contains incredible tax incentives to increase exports. Find out how many successful U.S. companies use these tax incentives to reduce U.S. taxes on their export sales by up to 75%.

> **Take advantage of the U.S. foreign tax credit.** Improper use could cause your taxes on overseas transactions to double. Learn what astute U.S. companies do to keep their U.S. and foreign tax liabilities low, thus increasing their bottom lines and stock prices.

WHAT INDUSTRY AND ACADEMIC LEADERS SAY ABOUT THIS BOOK...

"Our members will find this book easily readable and extremely valuable."

Senator Bill Campbell,
President Emeritus, California Manufacturers' Association

"The most useful treatment of international issues for companies...Its tight logic, clarity and brevity recommend it for use in business schools and poly sci departments."

Stephen S. Cohen,
Professor, Co-Director Berkeley Roundtable on the International Economy
(BRIE), University of California, Berkeley

"Absolutely superb. Both the legislature and business community will benefit greatly from this book."

Marz Garcia,
Former California State Senator from Silicon Valley

"This book covers many of the most important issues common to high tech and biotech companies. I strongly recommend it for biotech management teams."

Brad Goodwin,
Vice President-Finance, Genentech, Inc.

"This very readable book lays out for local government and industry representatives a host of valuable comments about companies in the Silicon Valley and elsewhere. It is an extremely valuable reference guide to have on your desk."

Carl Guardino,
President and Chief Executive Officer, Silicon Valley Manufacturing Group

"This book is ideal for our members, especially our smaller to middle size companies who are starting to sell in overseas markets and are trying to maximize their export sales, cash flow and bottom line profits."

Victoria Hadfield,
Vice President, Public Policy, Semiconductor Equipment
and Materials International, Inc.

"Professional service firms of any size will find this book essential in understanding the major sales, cash flow and bottom line public policy considerations faced by high tech and other U.S. companies."

James H. Henry,
Partner, Price Waterhouse LLP

"A very useful book for biotech executives who want to know how government policies impact their bottom lines."

John Krstenansky, Ph.D.,
Vice President, Research and Development, EnzyMed, Inc.

"Anyone involved in the regulatory aspects of a company in Silicon Valley, which means taxes, export licensing, export administration, customs or anything else dealing from the CFO on down, should have a copy of Cliff's book on his shelf for reference purposes and general guidance."

Larry R. Langdon,
Vice President, Tax, Licensing & Customs, Hewlett-Packard Company

"Its easy-to-read strong bottom line orientation makes **Beyond High Tech Survival** a necessity for company executives trying to compete in the U.S. and abroad."

Tom Meredith,
Senior Vice President/Chief Financial Officer, Dell Computer Corporation

"Business students preparing themselves for the world of high tech industries will find this book helpful in heightening their awareness of the many public policy issues that impact business strategies."

Annette Nellen,
Tax Professor, San Jose State University

"Cliff Jernigan takes complicated tax concepts and puts them in understandable English. They are the most readable tax explanations I have ever seen."

Robert H. Perlman,
Vice President, Tax, Licensing & Customs, Intel Corporation

"Cliff Jernigan makes complicated trade subjects readily understandable, a must read for any international business executive."

Clyde V. Prestowitz, Jr.,
President, Economic Strategy Institute, Washington, D.C.;
author of Trading Places and former U.S. trade negotiator

"Every member of high technology management teams should read this book."

Tom Proulx,
President and Chief Executive Officer, Netpulse
Communications, Inc.; author of Quicken and co-founder of Intuit

"A very worthwhile book on public policy issues in high technology.
CEOs and their management teams will find it very valuable."

W.J. Sanders III,
Chairman and Chief Executive Officer, Advanced Mirco Devices

"Cliff Jernigan's book is an outstanding addition to the business community and especially to the business school programs at our universities."

Thomas M. Stauffer,
President, Golden Gate University

"A very valuable reference guide for public affairs and corporate communication specialists who want to have a quick, easy-to-read synopsis of the
key public policy issues impacting corporate America today."

Bill Stotesbery,
GTT Communications, Inc.

"This book will be of keen interest to California companies as they try to
cope with burdensome public policy issues."

Kirk West,
*Chairman, Goddard Claussen First Tuesday
and former President, California Chamber of Commerce*

"This book hits the mark in helping companies reduce their costs from
frivolous lawsuits."

Don Wolfe,
*Silicon Valley Citizens Against Lawsuit Abuse,
Mayor, City of Saratoga, California,
Co-Host and Producer of the T.V. program "Issues Today"*

"Cliff Jernigan provides a solid overview of the primary public policy
issues affecting the competitiveness of high technology industries in the
United States today. This book will be of value to anyone working in—or
interested in—this sector."

Alan Wm. Wolff,
*Managing Partner, Dewey Ballantine Law Firm, Washington, D.C., and
former U.S. Trade Ambassador*

BEYOND
HIGH
TECH
SURVIVAL

turning government policy into
international profits

by cliff jernigan

BEYOND HIGH TECH SURVIVAL

turning government policy into international profits

by cliff jernigan

3 2280 00683 9955

Published by
OLIVE HILL LANE PRESS
2995 Woodside Road, Suite 400
Woodside, CA 94062 U.S.A.

ISBN 0-9655769-1-4

ACKNOWLEDGMENTS

Appreciation is owed to the following individuals for their valuable insights:

Joseph Acayan—Advanced Micro Devices

Doug Andrey—Semiconductor Industry Association

Ben Anixter—Advanced Micro Devices

William Archey—American Electronics Association

Chuck Ballard—Advanced Micro Devices

Jim Brandes—Advanced Micro Devices

Mary Bradley—City of Sunnyvale

Gary Burke—NASDAQ

LaJune Bush—LaJune Bush and Associates

Bonnie Byers—Hale & Dorr

Benjamin Byrd—Polaroid

Bob Call—RJC and Associates

Teresa Casazza—American Electronics Association

Dana Chiodo—Roan & Autry

Leslee Coleman—Silicon Valley Manufacturing Group

Reed Content—Advanced Micro Devices

Anne Craib—Semiconductor Industry Association

Osmond Crosby—Internet consultant

Karen Davis—City of Sunnyvale

Kevin Dempsey—Dewey Ballantine, LLP

Peggy Duxbury—Environmental consultant

Sue Eckert—Institute for International Economics

Chris Elias—Silicon Valley Manufacturing Group

Matthew Frank—Biotechnology consultant

Katherine Friess—Black, Kelly, Scruggs & Healey

Virginia Gates—Price Waterhouse LLP

Jim Glaze—Semiconductor Industry Association

Carl Guardino—Silicon Valley Manufacturing Group

Victoria Hadfield—Semiconductor Equipment and
Materials International, Inc.

William Haerle—MCI

Daryl Hatano—Semiconductor Industry Association

Mike Hawkins—Siemens

Brenda Hendricksen—Advanced Micro Devices

Gunnar Hurting III—Incites Ventures

Gil Kaplan—Hale & Dorr

Patricia Ketchum—Advanced Micro Devices

Larry Langdon—Hewlett-Packard

Brett Layton—Advanced Micro Devices

Rona Layton—Sims & Layton

Norma Lozano—Advanced Micro Devices

Jason Mahler—Office of Congresswoman Zoe Lofgren

Fred Main—California Chamber of Commerce

Diane Matulich—Advanced Micro Devices

Larry McCarthy—California Taxpayers Association

Mark McConnell—Hogan & Hartson, LLP

Don McIntosh—Advanced Micro Devices

Walter Moore—Genentech

Larry Mowrer—Advanced Micro Devices

Jeff Munk—Hogan & Hartson, LLP

Mary Murphy—Consultant

Annette Nellen—San Jose State University

Robert Perlman—Intel Corporation

Clyde Prestowitz—Economic Strategy Institute

Richard Previte—Advanced Micro Devices

Tim Ransdell—California Institute

Tom Reilly—Advanced Micro Devices

John Richardson—European Commission, Washington D.C.

Richard Roddy—Advanced Micro Devices

George Scalise—Semiconductor Industry Association

Jill Scoby—Advanced Micro Devices

Paul Snyder—Public Strategies Washington

Mary Beth Sullivan—California Institute

Vince Tortolano—Advanced Micro Devices

JoEllen Urban—U.S. Patent and Trademark Office, Washington D.C.

Theresa Wilson—Community affairs consultant

Don Wolfe—Silicon Valley Citizens Against Lawsuit Abuse

Mike Woollems—Advanced Micro Devices

and all of my MBA Students at Golden Gate University

Appreciation is also owed to the following for their valuable insights:

Copy editing by Christine Hopf-Lovette

Copy proofing by Jeri Burdick

Marketing consultation by Dina Camamis

Book and cover design by Emily Spoon

Printing coordination by Rob Robinson

Printing by Far Western Graphics

TABLE OF CONTENTS

Ten Government Policies That Can
Enhance Your International Profits 2

What Industry and Academic Leaders Say
About This Book 4

Acknowledgments 11

Table of Contents 15

Table of Tables 22

About the Author 23

Preface 25

Introduction 27

Chapter One – International Trade Issues 33

Aggressively Opening Foreign Markets: Section 301 34

Section 301: Actions Against Major High
Technology Countries 37

Super 301: The Impact on U.S. Companies 37

U.S.- Japan Semiconductor Trade Agreement:
An Example of the Aggressive Use of Section 301 38

Preventing Unfair Foreign Pricing in Your Market:
Anti-dumping Law and Countervailing Duties 40

    Dumping 40

    Anti-dumping Law 40

    Countervailing Duties 42

    Dumping of High Technology Products into
    the U.S. Market, including Dumping Margins 43

        Table 1.1 Dumping Margins (Percentages)
        by Company – Japan 43

        Table 1.2 Dumping Margins (Percentages)
        by Company – Korea 44

        Table 1.3 Dumping Margins (Percentages)
        by Company – Taiwan 44

    Short Supply Provisions 44

Reducing and Avoiding Tariffs in the U.S.
and Foreign Countries 45

Coping with the Export Control Laws 48

Handling Encryption Products and Export Controls 51

Making the World Trade Organization (WTO)
Rules Work for You 53

Competing in the European Union (EU) 55

Anticipating the European Union's New Currency
— the Euro 56

        Table 1.4 Time Table for the Euro 58

        Table 1.5 Euro Qualifying Countries 59

Taking Advantage of the North American
Free Trade Agreement (NAFTA) 60

Fending Off Foreign Company Acquisitions of Strategic
U.S. Companies: Using the Exon-Florio Act 61

Avoiding the Foreign Corrupt Practices Act 63

Dealing with China and Most Favored Nation Status 64

Understanding China's Prospects of Joining
the World Trade Organization (WTO) 66

Chapter Two – Company Lawsuit Exposure Issues 69

Handling Lawsuit Abuse 70

Defending Against Frivolous Securities Lawsuits 72

Anatomy of a High Technology Securities Lawsuit 74

Avoiding Product Liability and Punitive
Damage Awards 75

Working with the Food and Drug Administration 77

Table 2.1 Product Development Chart 78

Complying with the Clean Air Act 79

Understanding the Clean Water Act 80

Dealing with Leaking Underground Storage Tanks 82

Encouraging Environmental Responsibility
Through Tax Policy: A New Idea
—The Carrot and Stick Approach 84

Chapter Three – Antitrust and Intellectual Property Issues 87

Antitrust 88

Utilizing the Antitrust Laws for Research and
Development and Production Joint Ventures
with Your Competitors 88

Busting Monopolies: Making Competition Fair 90

Working with the Antitrust Laws of the U.S. and
European Union: A Comparison 93

Taking Advantage of Telecommunications Reform 94

Intellectual Property 96

Coping with Submarine Patents 96

Keeping up with the Internet and the
Copyright Laws 98

Aggressively Using the Law to Fight Intellectual
Property Pirates 99

Recognizing Other Types of Piracy 100

Stopping the Misappropriation of Trade Secrets 101

Protecting Semiconductor Mask Works from
Piracy: Semiconductor Chip Protection Act 102

Collaborating with the U.S. Government to
Protect Your Intellectual Property: "Special 301" 104

Table 3.1 "Special 301" Annual Review (April 30, 1998) . . . 105

Using the Intellectual Property Statutory
Protection Periods 107

Chapter Four – Federal Taxation 109

Maximizing Your Cash Flow Through
Accelerated Depreciation 110

Aggressively Using the Research and
Experimentation Tax Credit (Including State Analysis) 111

Utilizing the Orphan Drug Tax Credit 114

Understanding Employee vs. Independent
Contractor Tax Issues 115

Planning for Capital Gains 116

Avoiding the Alternative Minimum Tax (AMT) 117

Anticipating Alternative Tax Systems 119

Chapter Five – International Taxation 123

Taking Full Advantage of the Foreign Tax Credit:
General Principles 124

Enhancing Your Foreign Tax Credit: The
Treatment of Foreign Source Income and Expenses 127

Lowering Your Tax Rate on Exports: Using
the Foreign Sales Corporation (FSC) 128

Table 5.1 Most Common Methods of Calculating
FSC Tax Benefit 129

Aggressively Deferring Taxes on Offshore Operations:
The Concepts of Controlled Foreign Corporations
and Subpart F Income 130

Selling to Your Overseas Subsidiaries:
Being Aware of Transfer Pricing Issues 132

Manufacturing in U.S. Possessions: The U.S.
Possessions Income Tax Credit: (Section 936) 135

Taking Advantage of Tax Treaties 136

**Chapter Six – Employee
Compensation Issues** 139

Helping Provide for Your Employees' Retirement:
Profit Sharing, 401(k), and Pension Plans 140

Understanding Stock Purchase Plans 141

Incentivizing Your Key Employees:
Stock Option Plans 142

Table 6.1 Income Tax Consequences of
NSOs and ISOs for Employee and Company 143

Table 6.2 Tax Consequences of Exercise
of Stock Option: Hypothetical Real Life Example 144

Using Cafeteria Plans 145

Medical Reimbursement Plans 145

Child Care or Dependent Care Plans 145

Using Employee Tuition Reimbursement Plans:
Section 127 146

Chapter Seven – Workplace Issues 149

Complying with the Affirmative Action Laws 150

Avoiding Sexual Harassment Suits 151

Avoiding Wrongful Termination Suits 152

Assuring a Qualified Workforce: Immigration
and H-1B Visas 154

Utilizing Quality Improvement Teams 156

Lobbying for Flex-Time 157

Improving the Workers' Compensation System 158

Promoting Good Ergonomics for Your Employees 158

Chapter Eight – State and Local Issues 161

Helping Ease the Traffic Gridlock for Your Employees 162

Hiring and Developing an Educated Workforce 164

Supporting Affordable Housing for Your Employees 165

Lowering Your Utility Costs 166

Cutting through the Permitting Bureaucracy 167

Reducing Health, Safety and Environmental Dangers 169

Chapter Nine – State and Local Taxation 173

Lowering Your State Income Tax Bill 174

Working with the Unitary Tax System 175

Table 9.1 State Corporation Income Taxes:
A Comparison of Major State Taxes as
of March 31,1998 178

Minimizing Your Company's Property Taxes 180

Table 9.2 Personal Property Taxes:
A Comparison of Major States as of March 31,1998 183

Using the Sales and Use Tax Rules to Your Advantage 185

Table 9.3 Sales and Use Taxes: A Comparison of
Major States as of March 31,1998 187

Containing Other Local Taxes and Fees 189

Taxing the Internet—Good or Bad Idea? 192

Chapter Ten – Economic Development and Plant Site Selection 197

Understanding State Economic Development Programs 198

Selecting Your Next Capital-Intensive
Domestic Plant Site Location 199

Table 10.1 Factors to Consider in Selecting a
Domestic Plant Site Location (Worksheet) 200

A Tale of Two Cities: Comparing Austin, Texas
and Sunnyvale, California for a
Capital-Intensive High Technology Investment 202

Table 10.2 A Comparison of Austin and
Sunnyvale for a Capital-Intensive High
Technology Investment 203

Selecting Your Next Software Company Site Location 204

Table 10.3 A Comparison of Eleven Software Company
Site Locations in the U.S. 205

Selecting Your Next International Plant Site Location 206

Table 10.4 A Comparison of Twenty International
Plant Site Locations 208

Chapter Eleven – Political and Charitable Activities 211

Contributing to Politicians 212

Setting Up In-House Government Affairs
Departments 213

Giving to Charities 215

Writing to Public Officials 217

Meeting Public Officials 217

Chapter Twelve – Summary 219

Index 221

Table 1.1 Dumping Margins (Percentages)
by Company – Japan 43

Table 1.2 Dumping Margins (Percentages)
by Company – Korea 44

Table 1.3 Dumping Margins (Percentages)
by Company – Taiwan 44

Table 1.4 Time Table for the Euro 58

Table 1.5 Euro Qualifying Countries 59

Table 2.1 Product Development Chart 78

Table 3.1 "Special 301" Annual Review (April 30, 1998) 105

Table 5.1 Most Common Methods of Calculating
FSC Tax Benefit 129

Table 6.1 Income Tax Consequences of
NSOs and ISOs for Employee and Company 143

Table 6.2 Tax Consequences of Exercise
of Stock Option: Hypothetical Real Life Example 144

Table 9.1 State Corporation Income Taxes:
A Comparison of Major State Taxes as
of March 31,1998 178

Table 9.2 Personal Property Taxes:
A Comparison of Major States as of March 31,1998 183

Table 9.3 Sales and Use Taxes: A Comparison of
Major States as of March 31,1998 187

Table 10.1 Factors to Consider in Selecting a
Domestic Plant Site Location (Worksheet) 200

Table 10.2 A Comparison of Austin and
Sunnyvale for a Capital-Intensive High
Technology Investment 203

Table 10.3 A Comparison of Eleven Software Company
Site Locations in the U.S. 205

Table 10.4 A Comparison of Twenty International
Plant Site Locations 208

ABOUT THE AUTHOR

A California native, Cliff Jernigan holds a bachelor's degree in history, with a minor in English, from the University of Oregon, a law degree (J.D.) from the University of California's Hastings College of the Law, an advanced tax law degree (LL.M) from the New York University School of Law, plus numerous courses in accounting and management. He is a long-time member of the California Bar Association.

Jernigan has been associated with a well-known New York international law firm and has held corporate tax counsel positions at Bank of America, Stauffer Chemical Company, Castle and Cooke (now Dole Foods), and Advanced Micro Devices (AMD). He is currently Director of Worldwide Government Affairs for AMD. He also has served as corporate secretary for Bank of America's worldwide leasing subsidiary, B.A. Leasing Corporation.

Cliff Jernigan has been on the leading edge of major trade and intellectual property issues with the European Union, Japan, China, Korea, and Russia. He has participated in corporate tax research and planning and plant-site decision making for his corporate employers in the U.S., Europe, Asia, and Latin America.

A natural teacher, Jernigan has been a part-time adjunct professor at Golden Gate University over the past 16 years. His classes have included a law school teaching assignment in international taxation and MBA courses on the legal aspects of international business transactions, the impact of government on the high technology community, and government and the legal environment.

This book is the culmination of his broad education and experiences in the business community and reflects his deep understanding and insight into tax, trade, and other public policy business issues. This book is a follow-up to his first published book in 1996 titled *High Tech Survival—The Impact of Government on High Tech and BioTech Companies,* also published by Olive Hill Lane Press.

PREFACE

This edition of *Beyond High Tech Survival* was undertaken to help the business community (primarily company management) get a better handle on the major public policy issues impacting the sales, cash flow and profits of their companies. This book is meant to be your survival guide as you come upon government policies which may mean opportunities or disaster. Elected officials, government affairs representatives, government employees, law firms, accounting firms, venture capitalists, corporate board members, business schools, political science and economics departments will also benefit.

It is my hope that this book will someday be the standard public policy text for the business community, appearing on every executive's desk, much like the Strunk & White's *Elements of Style* appears on the desk of every writer.

Beyond High Tech Survival is for the lay reader. The intent of the book is to present public policy issues in a concise and accurate way so that the reader can gain a broad general understanding of each subject matter with only a few minutes of reading.

Included in the materials are important evolving issues and trends. This book is designed to be even more reader friendly than its predecessor: *High Tech Survival—The Impact of Government on High Tech and BioTech Companies.* While it is an extremely useful book for the high tech community, its scope applies to all of corporate America.

If the reader wishes greater knowledge of any issue areas covered in this book, he or she may obtain more in-depth materials from public and law school libraries, the Internet and specialists in the field.

The reader should not rely on statements in the book for legal or tax advice; you should consult with a legal or tax advisor as appropriate.

I believe you will find the materials worthwhile.

INTRODUCTION

We are in troubling waters. Many of the economic trends are disturbing. Perhaps as many as one-third of our companies will not survive the next five years. Companies will need to be nimble.

Your company may have leading edge research and development capability, the lowest manufacturing costs, an insatiable market demand, and the best marketing and sales team in your industry, and yet unforeseen government policies could slash your sales and take your bottom line from black to red.

What are the major government policies facing U.S. corporations today? How can we use government policy to help our survival? More importantly how can we turn government policy into international profits?

At the time of this writing, the U.S. economy is doing well, inflation is under control, the stock market is reaching new highs, the dollar is strong, and the unemployment rate is low. And yet troubling signs are emerging for U.S. companies that will make it harder for them to sell into Asia and Europe and compete in the U.S.

Asia is having problems. Japan's economy is in recession. Many Southeast Asian economies are having debt problems, their currencies are devaluing relative to the dollar, and unemployment is increasing.

Europe is experiencing some positive signs. Unemployment, while high, is declining. Many of the reforms of the European Union are paying off in increased efficiency and more competitive products. The new currency of the European Union—the euro—will make Europe still more competitive.

Three major populated countries—China, India and Russia —are starting to flex their muscle. Their internal markets are huge. Together these countries comprise one-third of the world's population. India, for example, has a middle class the size of the population of the U.S. These are difficult countries in which to sell and to do business. Bureaucracy and corruption are a problem. If they can overcome many of their inefficiencies, they will become economic powerhouses.

How will these world conditions impact U.S. companies? The impact could be considerable. Our ability to sell into their markets may be in some jeopardy.

In the case of Japan and those countries experiencing financial problems—Korea, Indonesia and Thailand—their buying power for strong-dollar U.S. goods will be tested. With unemployment rising, there will be nationalistic pressures to buy from companies within their countries. Protectionist sentiment represented by tariff and non-tariff barriers is certain to appear. U.S. companies already are reporting sluggish sales to Asia because of its financial condition.

Looking to Europe, competition will be intense. With Europe's new leanness and competitive posture, U.S. companies will have a harder time winning sales.

China, India and Russia want more share of the economic pie. They have programs already in place to protect their local industry from competition while supporting export programs to win greater world market share. U.S. companies will have difficulty selling into these protectionist markets.

At the same time, some U.S. companies are complaining of an increase in unfair foreign competition in the U.S. market. They claim the Asian countries are trying to increase exports as a way to solve their high unemployment problems. With their cheaper currencies and aggressive export pricing (some claim unfair pricing), Asian companies will put severe competitive pressures on U.S. companies in the U.S. market and elsewhere.

Europe should sell more products into the U.S. market as well. European companies will gain U.S. market share because of their greater efficiency and more competitive product positions.

China, India and Russia dream of harnessing their great science and math abilities into productive enterprises that could utilize their cheap labor rates to increase their market share, again perhaps through unfair pricing of their goods in the U.S.

While the competitive pressures on U.S. companies could become extreme, it is not always clear that the U.S. government recognizes these pressures. Nor does it appear ready to help solve this problem through various public policy responses. One cogent example is in the high tech area where U.S. policy on exports of sensitive products, particularly encryption and other leading edge items, seems at odds with keeping U.S. companies competitive with companies in Europe and Asia.

Our legal system often works against those aspects of business in which U.S. companies need to excel, i.e. our intellectual property and our efficiency. For example, our legal system may not adequately protect our inventions against those who would illegally appropriate and copy them, and it may curb through antitrust activity those cost-cutting efficiencies that may make U.S. companies more competitive through merger activity.

Cost pressures do not always arise from competition. In the U.S., companies are experiencing ever greater lawsuit activity —some say abuse—in the areas of securities, product liability, environmental, health and safety and workplace issues. These lawsuits drain our companies of their cash and management resource time.

Tax policy adds another unpleasant cost pressure. U.S., state and local taxation favors consumption over investment and penalizes research and development and capital-intensive companies. We do not enjoy the tax deferred cash flow benefits of our international competitors as we compete for sales in overseas markets.

We are trying to move faster to meet these competitive challenges. But quality of life issues are posing a problem. We work longer hours today. Two wage earner families are the norm. Our incomes are burdened by heavy commuting costs, expensive housing, work training costs and onerous income, social security, sales and use and property taxes. Sometimes it seems we are going backwards.

It is in this context—the worldwide public policy landscape —that this book was written. As we approach the twenty-first century, how can we use government policy to improve our quality of life and our standard of living? How can we use government policy as a tool to increase the competitiveness of our companies?

In the next twelve chapters of this book, encompassing nearly one hundred major government policy issue areas, I attempt to provide you with perspective to meet these challenges. Think of *Beyond High Tech Survival* as your public policy survival guide.

EUROPEAN UNION EURO

U.S. DOLLAR

JAPANESE YEN

RUSSIAN RUBLE

C H A P T E R   O N E

INTERNATIONAL TRADE ISSUES

KEY TOPICS

> *Aggressively Opening Foreign Markets: Section 301*

> *Preventing Unfair Foreign Company Pricing in Your Market: Anti-dumping Law and Countervailing Duties*

> *Reducing and Avoiding Tariffs in the U.S. and Foreign Countries*

> *Coping with the Export Control Laws*

> *Handling Encryption Products and Export Controls*

> *Making the World Trade Organization Rules Work for You*

> *Competing in the European Union*

> *Anticipating the European Union's New Currency—the Euro*

> *Taking Advantage of the North American Free Trade Agreement*

> *Fending Off Foreign Company Acquisitions of Strategic U.S. Companies: Using the Exon-Florio Act*

> *Avoiding the Foreign Corrupt Practices Act*

> *Dealing with China and "Most Favored Nation" Status*

> *Understanding China's Prospects of Joining the World Trade Organization*

INTERNATIONAL TRADE ISSUES

U.S. companies must be competitive in a global market. This chapter discusses what to do when foreign countries block access to their markets, when foreign competitors engage in injurious dumping practices, and when high tariffs and burdensome export controls make you uncompetitive. The chapter also discusses the North American Free Trade Agreement (NAFTA), the World Trade Organization (WTO), the European Union (EU), the Exon-Florio Act, the Foreign Corrupt Practices Act, China "Most Favored Nation" tariff status, and China's prospects of joining the World Trade Organization.

Aggressively Opening
Foreign Markets: Section 301

U.S. companies doing business abroad often complain about barriers to entry. Sometimes the barriers consist of high tariff rates. More commonly, they are subtle and consist of non-tariff barriers disguised as product standards, product testing procedures, product certifications, safety and environmental standards, bureaucratic and slow customs entry, local government or local industry majority ownership, lack of convertibility of local currency into U.S. dollars, and outright import bans.

Section 301 of the Trade Act of 1974 is a U.S. trade law provision that can be used to open up a foreign market. Section 301 authorizes U.S. retaliation against a foreign country which is in breach of a U.S. trade agreement or is engaging in unjustifiable, unreasonable or discriminatory conduct that burdens or restricts U.S. commerce.

A company initiates a Section 301 action by filing a

complaint with the United States Trade Representative (USTR), who investigates the complaint against the country and determines what actions need to be taken to remedy the situation. The USTR can also initiate a Section 301 proceeding on its own.

Among the remedies the USTR may take are retaliatory tariffs, quotas, and import bans on goods of the country. Usually, however, the USTR will try to avoid such retaliation and push instead for positive market-opening efforts, such as negotiating with the country to drop certain offensive standards or other non-tariff barriers.

To be successful in your Section 301 case, you will need to hire a competent law firm specializing in trade matters. Many of these firms have offices in Washington, D.C., where they are able to work closely with the USTR.

Perhaps the most famous Section 301 case is one that involved the U.S. semiconductor industry and its efforts to open up the Japanese market. The industry convinced the USTR that its sales in Japan were out of line with its sales to other regions of the world. The Section 301 complaint, filed in 1985, showed that the U.S. industry could not exceed a 10% market share in Japan over a 1975-1985 period despite numerous sales and marketing efforts. At the time, its market share was 75-80% in the U.S. and about 50% in Europe and Southeast Asia.

The USTR negotiated with Japan to open its market for semiconductors, setting a 20% foreign market share goal to be achieved by 1991. The goal was not met. In 1991, a second agreement was entered into to increase foreign market share to at least 20% by the end of 1992, with gradual and steady progress thereafter. By the end of 1995, foreign share was about 30%, of which the U.S. share was 18%.

One could conclude this U.S.-Japan Section 301 case was effective in helping open the Japanese market. For more information on the U.S.-Japan Semiconductor Agreement, see pages 38 and 39.

Section 301 is now commonly used by U.S. industry as a tool to gain foreign market access. The American Electronics Association (AEA) recently filed a Section 301 case to open up the Japanese market in computers and telecommunications. Kodak filed a Section 301 case against Japan relating to film for cameras.

With the conclusion of the Uruguay GATT Round in 1994, provisions in the Uruguay Round agreements require that Section 301 cases involving GATT (now WTO) members be sent to the WTO for dispute resolution. However, to the extent there is no WTO trade coverage of an area, or when the matter involves a country not a member of the WTO, Section 301 remains unchanged.

For example, many countries, including China, Taiwan, Russia, some Commonwealth of Independent States Countries (the former Soviet Union), and some Eastern European countries are not members of the WTO. They will still be subject to Section 301. Practices outside of the WTO, and thus still subject to a Section 301 case, include such areas as export targeting, collusion, and bribery.

WTO is inadequate in its coverage of areas such as patent protection for products under development and forced technology transfers. The latter category comes about where the foreign government requires a U.S. company selling into that country to license its technology to a local firm or to the foreign government. Section 301 still applies in these situations.

TAIWANESE DOLLAR

Section 301: Actions Against Major High Technology Countries

Many of the cases involving Section 301 have been in the high technology sector. Some of these cases are described below.

COUNTRY	UNFAIR PRACTICE
Brazil	Lack of intellectual property protections generally, lack of copyright protection for computer software, and lack of patent protection for pharmaceuticals.
China	Lack of intellectual property laws and market access.
European Union	Unfair satellite launching subsidies.
India	Lack of intellectual property laws.
Japan	Lack of semiconductor market access, market access for satellites, super-computers, and film for cameras.
Taiwan	Lack of copyright protection.
Thailand	Lack of intellectual property protection for books, records, movies, and pharmaceuticals.

Super 301: The Impact on U.S. Companies

Super 301 is an extension of Section 301. Part of the Omnibus Trade and Competitiveness Act of 1988, it offers U.S. companies improved tools to combat unfair trade practices. Requiring yearly extensions, it was extended in 1989 and 1990, but lapsed until 1994, when it was extended again and has continued to until September 30, 1998.

Super 301 requires the USTR to identify trade offenders. However, Super 301 is different in 1995-1998 from the Super 301 of 1988-1990. In 1989, the foreign countries listed were Japan, Brazil and India. Now, foreign trade barriers are identified under Super 301, but foreign countries are not.

Examples of foreign trade barriers are procurement restraints on purchases of U.S. supercomputers and space satellites, import bans, product tariff classification, licensing controls, barriers to foreign investment, foreign insurance, auto imports and agriculture.

Super 301 has been a thorn in the side of many foreign competitors. Super 301 goes beyond Section 301 by actually permitting the USTR to identify practices with significant trade barriers by September 30 of each year. Once identified, the USTR can commence an investigation. If the investigation shows a trade barrier problem, USTR can initiate a Section 301 filing, with all the attendant possible trade sanctions Section 301 provides.

U.S.-Japan Semiconductor Trade Agreement: An Example of the Aggressive Use of Section 301

This agreement, first signed in 1986 and later extended in 1991, calls for Japanese efforts to open their market to U.S. and foreign semiconductor sales. The trade agreement arose from a time of U.S.-Japan trade friction when the U.S. (and foreign) penetration of the Japanese market accounted for less than 10% of the total market.

The U.S. Department of Commerce and International Trade Commission found that Japanese companies were dumping semiconductor memory devices (DRAMs and EPROMs) in the U.S. market and causing injury to the U.S. semiconductor industry.

The U.S. government pursued one DRAM dumping case to conclusion. An additional DRAM case and the EPROM cases, however, were suspended as part of an agreement to open the Japanese market. With the DRAM and EPROM cases suspended, the Japanese agreed to work toward a target (the U.S. industry and U.S. government claim a commitment) of 20% of the Japanese market.

The resulting market-opening agreement was entered into in 1986. The 1986 agreement required Japanese semiconductor companies and the Japanese government to monitor dumping practices. In addition, the Japanese companies were to individually publish cost-of-production levels under which they could not sell.

Since 1986, U.S. and foreign companies have gradually increased their share of the Japanese market. In the 1991 agreement, the 20% share agreement was replaced by a "best efforts" agreement to sell more chips in Japan using a "gradual and steady" progress test. Cost of production levels for DRAMs and EPROMs were removed, although some dumping monitoring was retained.

The 1991 agreement expired at the end of July, 1996. At the end of the last quarter of 1995, foreign share stood at about 30%. Of this, the U.S. share was 18% and other foreign countries' share was 12% (composed primarily of the Korean share).

Many believe the 1986 and 1991 trade agreements have helped build trust and understanding between the two countries and have reduced trade friction. Executives of the two industries now talk to each other on a regular basis. As of the end of 1996, foreign share was well in excess of 30%. The agreements have been working.

Effective August 1, 1996, the U.S. and Japanese semiconductor industries negotiated a new three-year agreement with no percentage target. The European Union and Korea have also joined in the agreement, and Taiwan will join in the near future. Other countries may also join.

The key provisions of the new agreement are continuation of existing cooperative activities between users and suppliers, as well as new cooperative activities among suppliers. These activities include international standards, design-ins, and environmental programs. The agreement also calls for a continuation of data collection, but at a lesser level.

DANISH KRONE

Preventing Unfair Foreign Pricing in Your Market:
Anti-dumping Law and Countervailing Duties

One of the quickest ways to be driven out of business is to be the victim of unfair foreign company pricing. This can take the form of foreign dumping of goods, unfair cost advantages of foreign competitors, or predatory pricing by such competitors. Many companies have lost millions of dollars of sales and profits from dumping and other unfair pricing situations.

What can you do when you are one of these victims?

Dumping

Dumping occurs when an exporter sells in a foreign market at a price below the price he charges in his home market or when he sells below his cost, and the imports injure producers in the market of importation.

For example, assume Corporation A from Country X sells its product in Country X for $5.00 per unit and in Country Y for $3.00 per unit, a price which causes financial injury to competing companies in Country Y. This would be dumping. In the alternative, assume Corporation A's cost of production is $4.00 per unit in Country X and it sells its product for $3.00 per unit in Country Y, thereby causing financial injury to the competing companies in Country Y. This also would be dumping.

Anti-dumping Law

The U.S. and most foreign countries prohibit dumping. They have put in place so-called anti-dumping laws that make it illegal to sell in an importing country below home market price or home market cost when such pricing injures a domestic industry.

Injury to a domestic industry could occur, for example, when the industry's profit margins drop substantially or when many in the industry, including the cost-efficient companies, report losses as a result of the dumping.

The penalty for such behavior is to impose upon the exporter anti-dumping duties designed to increase the exporter's selling price to the higher of his home market price or his cost of production (including a reasonable profit).

A dumping action is commenced by going before the International Trade Commission to determine if the imports have caused an injury to a domestic industry and before the U.S. Department of Commerce to determine the amount of dumping duties, if any, to be applied. You will need a competent law firm specializing in trade matters to handle your case.

The philosophical rationale for making dumping illegal is the protection of local producers from unfair competition. Through pricing below market or cost, it would be possible to put an industry in the import country out of business. Such a result occurred, for example, in the U.S. semiconductor industry when many U.S. companies were financially injured and had to drop out of the dynamic random-access memory (DRAM) business because of Japanese dumping.

In retrospect, the U.S. semiconductor industry might possibly have reduced its injury if it had been more vigilant about the dumping threat and had acted more quickly to file anti-dumping actions. Nonetheless, the industry learned lessons that will help it address future dumping issues.

One of the early warning signals was a worldwide build-up of semiconductor wafer fabrication facilities which would eventually lead to an over-capacity situation. Another was steep drops in prices. Still another signal was a weakening of the U.S. dollar versus the Japanese yen. The stronger yen made the Japanese cost structure more expensive, with the consequence that Japanese DRAMs would have to be sold below cost in the U.S. market for the Japanese companies to stay competitive.

The industry learned other lessons:

> The anti-dumping laws were not always an effective deterrent to dumping. Some of the Japanese companies dumping DRAMs were the same ones who had previously been found to be dumping other electronics products. Thus, the U.S. anti-dumping law did little to deter multiple dumping situations.

> Anti-dumping duties did not go to the aggrieved companies, or industry, but rather to the U.S. Treasury. Thus, the damaged U.S. industry was not able to help rebuild itself with dumping duties.

> Pursuit of a dumping case is time-consuming and very expensive (about one-half to one million dollars in legal fees). In fact, an average dumping case could take up to two years, equal to the life cycle of some products. Thus, once dumping violations are found, it may be too late to help the damaged manufacturer.

Despite these and other shortcomings, the anti-dumping laws represent one of the major trade weapons of any industry and help to prevent injurious and predatory pricing. While most U.S. industries have suffered from foreign dumping of goods, dumping in the high tech sector has been a particular problem.

A listing of the major high technology countries whose companies have violated U.S. anti-dumping laws and the dumping margin percentages necessary to stop the injuries to the injured industries is found in Table 1.1, Table 1.2, and Table 1.3.

Countervailing Duties

A parallel to the anti-dumping action is the countervailing duty action. Injury to a domestic industry must be found by the U.S. International Trade Commission. Countervailing duties, like anti-dumping duties, must be determined by the U.S. Department of Commerce.

The purpose of countervailing duties is to counter foreign government subsidies that make it possible for exporters from the foreign country to sell at prices which injure U.S. producers.

Examples of foreign government subsidies include manufacturing incentives, such as cash grants, free land and manufacturing facilities, tax holidays, research and development tax credits, and current write-offs of assets. Other examples include selling incentives, such as value added tax exemptions on export sales.

In practice, countervailing duties have been difficult to apply. It has been a challenge to prove domestic injury. Also, the U.S. government has been hesitant to accuse foreign governments of providing subsidies.

Dumping of High Technology Products into the U.S. Market, including Dumping Margins

Companies in three Asian countries have been found to have dumped high technology products into the U.S. since 1980. These companies are from Japan, Korea, and Taiwan.

The dumping margins shown in the table below represent the amount of duties the U.S. government found necessary to bring the selling price of the product to a non-dumped price. For example, if the cost to produce a product is $200 and its sale price in the dumped country market is $100, the dumping margin necessary to bring the price up to a non-dumped price is 100% ($100.00).

Table 1.1 Dumping Margins (Percentages) by Company — Japan

Product	Toshiba	Matsushita	NEC	Oki	Hitachi	Fujitsu	Mitsubishi	Sony
Microwave Ovens (1980)[1]	13							
High Capacity Pagers (1983)		110	70					
Cellular Mobile Phones (1985)		107	96	10	3			
DRAMs 256K (1986)[2]	50	107	107		20	74	107	
EPROMs (1986)[3]	60		188		85	103		
DRAMs 64K (1986)			23	35	12		13	
Color Picture Tubes (1988)	34	27			22		1	
3.5" Microdisks (1988)					28			51
Digital Readout Printers (1988)								39
Small Business Telephone Systems (1989)		137	179					
Supercomputers (1997)			454			173		

[1] Terminated at Petitioner's request after final determination
[2] Preliminary Determination/Suspension Agreement
[3] Suspension Agreement

Table 1.2 Dumping Margins (Percentages)
by Company — Korea

PRODUCT	GOLDSTAR	HYUNDAI
DRAMs 4 megabyte and above (1993)	5	11

Table 1.3 Dumping Margins (Percentages)
by Company — Taiwan

PRODUCT	ADVANCED MICROELECTRONICS	TEXAS INSTRUMENTS			OTHERS
		BIT	ACER	WINBOND	various up to:
SRAMs (1998)	113.85	113.85	113.85	102.88	93.87

In the case of Taiwan, the U.S. government had to address the issue of Taiwanese foundries that perform fabrication of SRAMs for fabless producers. (Fabless producers design and market under their own name, purchasing fabrication services from a foundry.) The government ruled that the margin rate for SRAMs will vary according to the design house, not according to the foundry. Rates assigned to foundries only apply to proprietary SRAMs that the foundry designs and markets under its own name.

Short Supply Provisions

Once anti-dumping duties have been imposed on a foreign company, the price of that company's imported goods would be expected to rise. With these rising import prices has developed an area of dispute among different interest groups in the U.S. over an issue called "short supply."

The argument advanced by some consumer industries is that if a U.S. industry cannot supply essentially all of the consumer industry's required product needs of a product subject to anti-dumping duties, then the anti-dumping duties should be removed.

Proponents of anti-dumping duties argue that:

> had there been no dumping, the U.S. producer industry would have been stronger and more able to meet the consumer industry's needs

> to remove anti-dumping duties where there is a short supply situation would reward the foreign company found to be dumping and encourage that company to dump even more. In other words, removal of the duty would permit renewed dumping, which would further weaken the U.S. producer industry's ability to meet the consumer industry's needs.

One of the strongest opponents of the short supply provision has been the U.S. steel industry.

This short supply issue has, on occasion, brought U.S. companies in the computer industry into heated conflict with companies in the semiconductor industry. To date, there has been little room for compromise on either side. The computer industry wants a short-supply exception. The semiconductor industry wants no exception. It remains to be seen whether a compromise position can ever be attained.

One possible area of compromise could be in cases where there is no U.S. supply to begin with, as in a case involving flat panel displays where it was debatable that a U.S. flat panel display industry existed. Short of this situation, it appears the parties will continue to be at odds over the short-supply issue.

Reducing and Avoiding Tariffs
in the U.S. and Foreign Countries

The area of tariffs is commonly referred to as duties. Tariffs are also referred to as customs. Throughout this topic, I use these terms interchangeably.

Tariff issues involve considerations of both public policy and bottom line profits. High tariffs in a country to which an exporter is selling can hurt his competitiveness. An absence of tariffs can improve an exporter's ability to compete.

Since 1947, with the establishment of the General Agreement on Tariffs and Trade (GATT) and in numerous succeeding meetings (called "rounds") culminating with the most recent Uruguay Round which began in 1986 and ended in 1994, the trend has been to lessen tariff rates.

In the electronics industry, for example, tariffs in some countries have dropped precipitously in the past few years, going to zero percent in some sectors of the industry.

Major customs policy involves the areas of product classification, valuation, country of origin and marking, and preferences and other tariff reductions.

> **Product Classification** is where products are separated into special categories, and a tariff rate is applied to each category. Classification is complex. Should a personal computer monitor, for example, be classified as a computer part eligible for zero tariff treatment, or as a TV which is subject to a 10% duty rate? European TV makers favor the higher rate. This type of issue currently poses a serious trade conflict between the U.S. and the EU.

> **Valuation** is the process of determining a product's value. Value is normally the purchase or market price, or "transaction value." When a transaction value cannot be determined, Customs may look at other methods to determine value.

> **Country of Origin** is generally the country where the product is manufactured, produced, grown or extracted. Origin becomes complicated when the product is assembled in one country with parts from other countries. For example, a computer assembled in Japan may contain sub-assemblies from Malaysia. Certain importing countries may view this as a product of Japan, but others will view it as a product of Malaysia. This creates a dilemma for the Japanese manufacturer as he needs to mark his computer with the correct country of origin. Importing countries require this information because duty rates are set not only on the value and classification of the material, but also its country of origin.

> **Preferences** relates to the granting of preferred tariff rates for products with origin from certain countries or regions.

In the U.S. this is called the Generalized System of Preferences (GSP), a provision that may permit certain developing countries to ship product to the U.S. duty free, provided they comply with certain regulations. These include meeting U.S. safety and labeling requirements, recognizing workers rights, protecting intellectual property rights, and not granting other nations more favorable duty rates than those granted to the U.S.

The Caribbean Basin Initiative (CBI) is just such a provision. Under the CBI, most Caribbean nations enjoy duty free entry into the U.S. for the bulk of the goods they manufacture and export.

U.S. possessions, such as Puerto Rico and Guam, are considered part of the U.S. and therefore also enjoy duty free entry into the U.S. for goods manufactured there.

Other methods of achieving tariff reductions include duty suspensions, duty drawback, and the use of free trade zones. Many companies have saved millions by using these tariff reduction methods.

> **Duty suspensions** are used most commonly in the EU. A duty suspension allows goods to enter into the EU duty free if the manufacturing capability for that product does not exist within the EU. This cost reduction assists EU importers by leveling the playing field with companies based in countries where the manufacturing capability does exist.

> **Drawback** is the refund of duty for goods imported and subsequently re-exported. There are two types of drawback: same condition drawback and manufacturing drawback.

Same condition drawback occurs when a product is brought into the U.S. and subsequently re-exported from the U.S. without losing its identity. The U.S. importer/exporter can apply for a 99% duty refund under terms of same condition drawback program.

Manufacturing drawback occurs when a foreign product is imported into the U.S., undergoes further manufacturing, and is then included in a new piece of equipment which is then exported out of the U.S. The product originally imported is exported out, but it loses its identity because it is now a part of the new product. The importer/exporter can receive a 99% duty refund.

> **Free trade zones** (FTZ) are areas usually, but not always, near international airports or ports. Products can be brought into a FTZ duty free, undergo manufacturing in the zone, and then be shipped out of the country. If the finished good is shipped out of the FTZ into the same country, then duty will be assessed at that time.

The area of customs administration and compliance is complex and arcane, with the possible imposition of civil and criminal penalties serving as a continuous threat to non-compliers. Needless to say, a company must have competent staff and outside consultants specializing in this area to achieve the lowest customs duties possible and without penalties.

GATT, which was changed to the World Trade Organization (WTO) in 1995, will be looking at many new customs issues. It is important that a company follow the activities at the WTO and protect its interests as appropriate. One area being considered is the further elimination of non-tariff barriers, such as standards and testing procedures, slow and bureaucratic customs administration and arbitrary application of laws and regulations.

Coping with the Export Control Laws

Most export controls are designed to prevent high technology goods from getting into the possession of unfriendly governments which could use such technology against the U.S. The U.S. also uses export controls to limit "short supply" products, such as some petroleum products and certain types of lumber, munitions (including some electronic devices), endangered species, and nuclear equipment, from shipment outside the U.S.

Modern U.S. export controls originated in the post World War II period. With the beginning of the Cold War with the Soviet Union, the U.S. needed to maintain its military advantage by ensuring its technical superiority. Although export controls were aimed at the Soviets, friendly nations to the USSR were also caught in the U.S. export control web. Unlike most countries, the U.S. also uses export controls as a "tool" of foreign policy. Recent examples include the embargoes of Iran, Iraq, and Libya in which virtually no goods can be exported to these countries.

The Department of Commerce and, to a lesser degree, the Department of State administer most U.S. export control programs. Both agencies issue licenses which may be reviewed by the Department of Defense, the Department of Energy, and the Arms Control and Disarmament Agency. Sometimes the Department of Commerce, whose charter is to promote U.S. exports, comes into conflict with the Departments of State and Defense, who put more emphasis on foreign policy and control of parts that have the potential of ending up in "munitions" devices.

Over the years, U.S. export controls have focused on software, semiconductors, semiconductor materials and equipment, computers, and telecommunication products. Industry groups and think tanks often assert U.S. export controls cost U.S. industry tens of billions of dollars of lost sales and hundreds of thousands of lost jobs. The government response has been that this is a "reasonable price" for national security.

U.S. industry has had to jump through many bureaucratic hoops because of restrictive U.S. export controls. Some products could not be exported without "individual validated licenses"—prior U.S. government approval. If a company had an effective export control policy, the Department of Commerce would occasionally grant a bulk license permitting the export and re-sale of specific products to a broad category of customers in a specific territory. Sometimes, when a U.S. seller demonstrated that a foreign competitive product was freely available outside of the U.S., the Department of Commerce would grant a license over the objections of the Department of Defense or other agencies. This concept is known as the "foreign availability" exception.

With the easing of the Cold War in 1989, some U.S. export controls have also been eased. The restrictions surrounding many exports of high technology products have been eliminated or reduced. However, in the area of weapons that could be used for mass destruction, export controls have been expanded to prevent the proliferation of such weapons. These newer proliferation controls are quite broad and cover even low-level technology.

Today, U.S. export controls typically apply to only the most advanced products. U.S. products are often forbidden to rogue countries and countries who may pose a national threat to the U.S. China, Russia and many of the former Soviet Union countries are the most prominent on this list.

A sleeper area is with respect to foreign nationals. Most companies do not know that export controls extend even to the hiring of some foreign nationals working in the U.S. Giving them access to controlled technology used in your U.S. manufacturing operations, for example, may be deemed to be a transfer of technology abroad.

This "deemed export" rule requires U.S. companies to obtain an export license if nationals of China, Iran, Russia and some other countries have access to controlled technology. With the recent nuclear tests in India and Pakistan, the "deemed export" rule may be extended to nationals of these countries as well.

Another area that still causes great concern for the electronics industry involves the control of encryption (data scrambling) devices and software. This includes some mass market software.

A group called the Computer Systems Policy Project (CSPP), composed of some of the major producers of computer systems, estimated in a 1995 study (Emerging Security Needs and U.S. Competitiveness: Impact of Export Controls on Cryptographic Technology) that the potential annual loss of revenues from export controls for the U.S. information industry could range from $30 to $60 billion dollars by the year 2000. And in a 1998 study the respected Washington, D.C. think-tank Economic Strategy Institute stated that the U.S. economy could lose anywhere from $35-96 billion between 1998-2002 if the U.S. government does not abandon its current policies on encryption.

The electronics industry has been putting considerable resources toward a solution to the encryption issue. An acceptable solution will permit the free export of products without compromising our national defense.

As we head into the twenty-first century, with the Cold War behind us, a new international export control agreement is in place to reduce the likelihood of sensitive goods falling into the wrong hands. Called the Wassenaar Arrangement, it has been adopted by many of the major countries, including the U.S. and Russia, to control the export of arms to rogue nations such as Iran, Iraq, Libya and North Korea. How well this new agreement will work remains to be seen.

Handling Encryption Products and Export Controls

Encryption is the science of using codes to protect electronic data from getting into the wrong hands. It is analogous to using a padlock to keep intruders out of your house. Using data scrambling technology, encryption is designed to protect electronic communications within your company, with other companies, and with your customers, distributors, and suppliers.

Many claim that the Global Information Infrastructure (GII) will not develop to its full potential until we can be sure that messages sent over it are private. If I want to sell this book over the Internet for a fee, I will need an encryption method to prevent freeloaders from downloading my full book without paying for it.

Sellers of many other types of property, including software, movies, and music, will have a need for similar protection. Banks, hospitals, telecommunications providers and multinational corporations will also require that their electronic communications be kept private by encryption techniques.

The U.S. electronics industry is the leading maker of encryption products. Because the industry sells a large percentage of its products overseas, it has a need to put encryption into its products to protect the privacy of its customers. These international users are demanding state-of-the art encryption technology to prevent intruders (including governments) from viewing and analyzing proprietary information.

Export control rules pose a great problem for those who want to sell encryption products overseas. State-of-the-art encryption products that can be sold in the U.S. are often not permitted by the export control laws to be sold overseas. In fact, the U.S. export control laws used to classify most encryption products as "munitions," placing them in roughly the same category as bombs. Beginning in 1997, the U.S. government removed encryption products from the "munitions" list.

The U.S. government recognizes that export control laws can adversely affect U.S. industry and has been seeking ways to relax the controls. At the heart of the issue for the government is the protection of our national security and law enforcement. Having the ability to tap into electronic data communications of criminals and terrorists is fundamental to the government's case. They want to keep state-of-the-art encryption technology out of the hands of such people and organizations.

The government position, promoted by the Department of Commerce, the FBI, and the National Security Agency, is to permit greatly relaxed export controls if the government can obtain the keys necessary to tap into electronic messages. This concept, called "key recovery," would create a duplicate of the software key used to read encrypted data and place control of the key in the hands of a trusted third party (either private or government). Upon a need, supported by a court-ordered subpoena, the third party would hand over the key to the agency making the criminal investigation.

The business community and many members of Congress respond that the key recovery system will not work. They believe it is too complex and that it would not be applied on a worldwide basis because other countries would not join in. They also argue that it would not bring in terrorists, drug dealers, and other lawbreakers who already have access to this technology and who would not register their keys with an escrow agent.

Perhaps the most important problem, according to the business community, is that the U.S. is not the only country making encryption products. Many countries producing state-of-the-art encryption products see stringent U.S. export controls as their way to win the encryption competition. The genie is already out of the bottle, the U.S. industry claims. Prohibiting U.S. companies from selling state-of-the-art encryption products overseas will only result in a loss of sales, jobs and the ability to continue building state-of-the-art encryption products.

Some U.S. companies are investing in foreign encryption product companies as a possible way to avoid the strict U.S. export control rules. They conclude that it would be ironic if our own U.S. government had to buy foreign encryption products to protect itself from criminals.

Recently the debate has begun to shift from one of law enforcement vs. industry to one of privacy, constitutional rights and crime prevention vs. Big Brother. This shift appears to be helping industry as the government quietly has been allowing state-of-the-art encryption technology to go overseas under some circumstances.

There will always be a need to balance economic interests with national security. Hiring of competent export control specialists and strong lobbying in Washington, D.C. will most likely lead to considerable government concessions and a much greater ability to export state-of-the-art encryption products.

Making the World Trade Organization (WTO) Rules Work For You

The WTO, formerly the General Agreement on Tariffs and Trade (GATT), was formed in Geneva in 1947 to reduce trade barriers among countries. In its present form, WTO has substantially reduced tariffs around the world and has helped the international trade of goods and services.

There have been several major meetings of the international community in a series of rounds, including the "Kennedy Round" of 1964-1967, the "Tokyo Round" of 1973-1979, and the "Uruguay Round" of 1986-1994. Each successive round has reduced worldwide trade barriers. The main features of the WTO are reflected by the articles of the Agreement. Some of the more important articles are as follows:

> **Article I** requires each country in the WTO (made up of over 100 members including the U.S., Canada, Mexico, the Western European countries, and many Asian and Latin countries) to grant "most favored nation" (MFN) status to all member countries, (i.e., tariff rates and other privileges granted by a member to any other member must also be granted to all of the other members).

> **Article VI** recognizes that dumping and unfair subsidies must not be allowed, and gives members recourse against other members.

> **Article XII** permits a member to restrict the entry of imports if its external financial position or balance of payments is threatened.

> **Article XIX** allows similar relief where imports threaten serious injury to domestic producers.

> **Article XXIV**(5) allows the formation of free trade areas, such as the European Union.

WTO serves as a complaint forum by member countries. WTO tribunals can cite member countries for violation of the WTO rules. WTO has been somewhat successful in encouraging its members to be more fair in their dealings.

For the business community, tariff relief, anti-dumping, and subsidy provision measures have been among the most helpful in permitting fair trade. Market access should also improve and tariff and non-tariff barriers further eliminated by concerted WTO action. As WTO rules are drafted and new rounds undertaken, the business community should be vigilant to protect their WTO gains.

Competing in
the European Union

For most companies with sales over $10 million a year, Europe represents rewarding business opportunities. Sales in the region generally account for about 25% of total sales for these companies. With such a large sales presence in a region growing to 400 million people, it behooves companies to become familiar with the rules of the European Union (EU).

The member states of the EU are Austria, Belgium, Britain, Denmark, Finland, France, Germany, Greece, Ireland, Italy, Luxembourg, Portugal, Spain, Sweden, and the Netherlands. Other European countries will probably join the EU in the future.

The EU is comprised of two legislative branches and one executive branch, all deriving their origin from the 1957 Treaty of Rome. One branch, the Council of Ministers, is akin to the United States Senate. Whereas each state in the U.S. is represented by two senators, each country composing the EU is represented by one minister. There are 15 ministers. Another branch, the Parliament, is akin to the United States House of Representatives, with country representation based on population.

The third branch of the EU is the European Commission (EC), located in Brussels. The EC is roughly akin to the Executive Branch of the U.S. Government. Its affairs are run by Commissioners from each of the member states. Each Commissioner is in charge of a specific area of EC law, called a Directorate General.

The EC is independent of the Council of Ministers and the Parliament, but major EC proposals, called Directives, must go before the Council and Parliament for their approval. Each of the Commissioners has a large staff, who are often referred to as Eurocrats.

Some of the Directorates General important to the business community include the following: Directorate General I (DGI) deals with external relations (e.g., there is a U.S. desk, a Japan desk, a Russia desk and a China desk) and trade policy (such as anti-dumping and countervailing duty law). DGIII deals with the area of industrial policy; DGIV administers competition (antitrust) policy; and DGXXI administers tariffs and rules of origin policy. The EC also negotiates trade treaties with other countries and is the main link with the World Trade Organization (WTO).

In the mid 1980s, the EU undertook an effort to harmonize the laws of the various member states. In a plan called Europe 1992, it embarked on about 300 directives in such areas as free movement of workers, free movement of self-employed persons to locate in any member country, free movement of capital, free provision of services, etc. These directives have created a single market in which a company can, with few exceptions, do business under a single set of rules throughout the EU.

For many U.S. businesses, dealing with the EU is an untapped opportunity. The EU seems more politically oriented than the U.S. Congress and Executive Branch. EU policies are often put into effect to allow latitude to serve the public interest rather than follow a strict reading of the law. U.S. companies and trade associations can potentially tap into EU policy makers and mold EU policy to their benefit.

To do so will require that you set up an office in Brussels or hire a competent law or accounting firm in Brussels to seek out EU opportunities for you. Many U.S. law and accounting firms have set up offices in Brussels to help you with such opportunities. The U.S. Mission to the EU, located in Brussels, can also be extremely helpful.

Anticipating the European Union's New Currency—the Euro

In 1999, the European Union (EU) will have a new currency—the euro. This is the monetary unit that will replace the individual country currencies for EU member countries, except for England, Denmark, Sweden and Greece, which will keep their present currencies for the time being.

The euro will have a big impact on the U.S. economy, U.S. corporations doing business in the U.S. and abroad, and executives traveling within Europe. The euro will challenge the U.S. dollar as the currency of choice in many areas, including international contracts, international financings, and a large eurobond market that will compete with the U.S. Treasury market.

European financial statements and employee wages and pay stubs will now be in euros. Bond issues and equity holdings also will be in euros.

U.S. industry doing business in the euro zones of the EU will face challenges and opportunities. The euro's impact will be felt in some of the following ways:

> **Companies doing business in Europe** will incur initial costs to convert to a euro-dominated marketplace.

> **U.S. interest rates** may rise. Germany may cause this increase as the dominant euro member country by working to contain inflation through higher interest rates. The higher euro interest rates may attract foreign investors away from the dollar. The U.S. may need to respond by increasing interest rates for its bonds, thus driving up U.S. interest rates. U.S. corporations could feel the impact in higher borrowing costs to finance research and development, new plants and equipment purchases. With the higher interest rates may also come lower profits and stock prices.

> **U.S. companies pricing** their products in a different currency to each country will now use the euro for each country. This will make differences in country pricing more obvious and may have the impact of driving down the price to that of the lowest country market price. For example, if I sell computers to Germany at a higher price than my sales price to France, these sale's price differences will not have the benefit of being hidden through the use of the German mark and the French franc. As a result, companies may see their profits eroding.

> **U.S. companies may save** on foreign exchange transaction and hedging costs as these companies will not have to make currency conversions for intra-Europe transactions.

> **U.S. executives traveling** through Europe will avoid the foreign exchange costs of converting from one country's currency to another each time they enter a new country.

The time table for the introduction of the euro is shown in the following chart:

Table 1.4 Time Table for the Euro

1992	Maastricht Treaty mandates that European Union (EU) countries joining the one currency Economic and Monetary Union (EMU) have budget deficits of less than three percent of gross domestic product and a national debt of less than 60 percent of gross domestic product by 1997.
1998	EU countries satisfying Maastricht Treaty mandates are eligible to join the EMU. Qualifying are Germany, France, Belgium, the Netherlands, Luxembourg, Ireland, Finland, Austria, Italy, Portugal and Spain. Greece will not currently qualify. For the time being England, Denmark and Sweden, all qualifying, will not join. Exchange rates between national currencies in the euro zone and the euro will be announced. The EU's central bank, and a new central bank president, will replace the European Monetary Institute.
1999	National currencies of EMU participants will now be set at a fixed rate of exchange with the euro. National currencies may continue to be used also, but pegged at the euro exchange rates.
2002	EMU member countries will introduce euro currency in exchange for their national currencies and discontinue these national currencies. The euro will now be the only currency.

EUROPEAN UNION EURO

The following table shows how the eleven qualifying member countries adopting the euro measure up to the 3% deficit and 60% debt tests:

Table 1.5 Euro Qualifying Countries

COUNTRY	DEFICIT/PLUS (% OF GDP)	DEBT (% OF GDP)
Austria	-2.3	64.7
Belguim	-1.7	118.1
Finland	+0.3	53.6
France	-2.9	58.1
Germany	-2.5	61.2
Ireland	+1.1	59.5
Italy	-2.5	118.1
Luxembourg	+1.0	7.1
Netherlands	-1.6	70.0
Portugal	-2.2	60.0
Spain	-2.2	67.4

European Commission Forecasts for 1998

Whether U.S. companies will be helped or hurt by the euro remains to be seen. With careful planning, a U.S. company very well could derive a benefit. Your finance department, especially your treasurer, will play a key role here.

DUTCH GUILDER

Taking Advantage of the
North American Free Trade Agreement (NAFTA)

In 1993, Canada, the U.S., and Mexico entered into an agreement for preferential trade treatment among the three countries. The agreement, called NAFTA, created a unified trading block of nearly 400 million persons.

Designed to provide economic security for businesses of the three countries and to lower consumer costs, NAFTA was in part a recognition of the competitive threat of other economic trading blocks, such as the European Union (EU) and the Association of Southeast Asian Nations (ASEAN), and in part a recognition that Mexico's economy could be strengthened by lowering costs to Mexican consumers and creating more jobs there.

Some of the main provisions of NAFTA include lower tariffs among the countries, elimination of many investment barriers to permit greater investment into each country, and freer movement of goods across each other's borders.

The value of NAFTA to the economies of the three countries has been a subject of vigorous debate. NAFTA proponents argue that lower tariffs and investment barriers will spur many new marketing opportunities and create jobs. Consumers will benefit because goods flowing among the countries will be produced by the most cost-effective and efficient work forces.

Canadian and U.S. NAFTA opponents argue that many jobs, primarily higher paying blue collar jobs in Canada and the U.S., will be lost as companies close down plants in Canada and the U.S. to build plants in Mexico. Mexican opponents argue that more efficient Canadian and U.S. companies will use their advanced manufacturing knowledge and technology to put local Mexican-owned companies out of business.

Has NAFTA been helpful to U.S. industry? Early data seem to indicate it has. In a 1996 study by researchers at the University of California at Los Angeles, increased imports to the U.S. resulted in losses over the past three years of 28,168 jobs while increased exports created 31,158 jobs.

The U.S. electronics industry is a positive case in point. U.S. computer companies, for example, have increased sales to Mexico. Such sales increases have been good for computer industry suppliers, such as semiconductor companies and their vendors (the equipment and materials companies). Cellular phone and pager producers are having similar success in selling to Mexico.

Recent tariff reductions among the three countries, primarily in the chemicals sector, should generate still more trade for U.S. companies.

There is a movement underway to expand NAFTA to many other countries in Latin America. The proposed Free Trade Area of the Americas could include 750 million people by 2005. Chile, Argentina and Brazil have a good chance to be among the first countries to be added.

It will be perhaps many years before a more-informed judgment can be made as to the wisdom of NAFTA and whether it has had a positive effect on dealings among the three countries. It seems evident, however, that these three countries will collectively be more competitive with other trading blocks, such as the EU and the ASEAN countries, and that this increased competitiveness will lead to new job creation.

Fending Off Foreign Company Acquisitions of Strategic U.S. Companies: *Using the Exon-Florio Act*

In 1988, the U.S. electronics industry became concerned about a lack of U.S. supply in certain subgroups of the industry. This concern involved the concept of the "food chain."

The idea of the food chain is that each area of the industry is critical to the well-being of the entire industry. A shortage of key chemicals or equipment used by the semiconductor industry could, for example, result in decreased production of the semiconductor chips used in personal computers and telecommunication products. This could in turn be harmful to U.S.

national defense and economic interests. From a national defense standpoint, if any one country has a virtual monopoly over a subgroup (especially if the country is unfriendly to the U.S.), U.S. national security could be imperiled.

It was with this background that the Exon-Florio Act was passed to prevent such foreign acquisitions that would damage national security. The Committee on Foreign Investment in the United States (CFIUS), already created in 1975, became the review board for foreign acquisitions.

One of the first applications of Exon-Florio was the proposed acquisition of Fairchild Semiconductor Corporation by Fujitsu of Japan. Part of Fairchild's business was in defense. CFIUS reviewed the proposal and determined there was a national security risk; however, the law was written in such a way that they could not stop the acquisition. Under pressure from U.S. and Japanese interests, however, Fujitsu gave up its acquisition effort. A short time later National Semiconductor Corporation purchased Fairchild.

CFIUS has made several hundred investigations since 1988, but only one case involved a blocked acquisition. This proposed acquisition related to a Chinese company's efforts to buy MAMCO Manufacturing, a Washington state aircraft manufacturer.

Exon-Florio is one of the few tools the U.S. government and industry can use to block a foreign acquisition. In the Fairchild-Fujitsu situation, lobbying of the U.S. Congress and Administration was fierce. However, many in the U.S. electronics industry feel Exon-Florio does not go far enough. It applies only to national defense situations, and industry executives argue it should be extended to economic security situations as well.

Congress has been reluctant to go this far. But if foreign competition becomes a scapegoat target for hurting U.S. competitiveness, it is likely Congress will again review the situation and possibly act to extend the law to include economic security. Be alert to lobby this situation if the opportunity arises.

Avoiding the
Foreign Corrupt Practices Act

An issue that spells extreme danger to corporations and their executives is the Foreign Corrupt Practices Act, passed in 1977. The law prohibits company and employee payments to foreign government officials or their intermediaries. Violation of this law can involve fines to the company and company employees, and, in extreme cases, jail terms.

While it may be an expensive and time-consuming exercise, companies must be aware of this issue and be sure that their policies and procedures reflect an understanding of the law. Many years ago, a U.S. company participated in a review of a transaction involving the mayor of a city in a third-world country. The mayor had asked the company to make a $100 contribution to him so he could buy uniforms for his town's Little League team. The company management team considering this request involved the chief financial officer, corporate controller, the general counsel, the tax director, and the company plant manager in the foreign city. It took about two hours of discussion, including perhaps $1000-$2000 of management time, to reach the conclusion that the gift of $100 was inappropriate.

The Foreign Corrupt Practices Act can be particularly galling to U.S. companies who lose foreign contracts to companies from other countries, such as France and Germany, and in most Asian and Latin American countries. But the U.S. has usually taken the moral high ground and the Foreign Corrupt Practices Act appears in no danger of being repealed.

There are some important exceptions to the Foreign Corrupt Practices Act. One exception relates to the concept of "facilitating payments" (also called "grease payments"). These payments do not violate the Foreign Corrupt Practices Act when they are made for the purpose of a routine government action.

An example of such a payment would be a payment to a customs official to speed up his review of customs documents so as to expedite shipping of the goods out of the country. The payment might have been necessitated, for example, when the only customs inspector at the airport engaged in slowdown tactics, such as threatening to go on vacation or getting sick, and the company, with the facilitating payment, was able to keep him on his job at the airport (in good health) in order to process the shipment.

Another major exception is a payment to a government official which would be viewed in the U.S. as an illegal bribe but, under the law of the country where the foreign official resides, is legal.

Legal review and advice is essential. Use of one of these exceptions should involve extreme caution. A company will not want to step over the line into an illegal area when it is intending to make a legal facilitating payment or a payment legal under the law of that country.

There have been attempts to modify the Foreign Corrupt Practices Act, the most recent of which was in 1988. Industry will push for even more modifications of the law where the Foreign Corrupt Practices Act becomes too onerous or hurts a company's competitive posture *vis a vis* foreign competitors who routinely make such payments and get away with it. The World Trade Organization may take up this issue in the near future.

Dealing with China and "Most Favored Nation" Status

The issue of China and "most favored nation" (MFN) status has arisen every year since 1980 and is always heatedly debated by the President and the Congress.

The source of the debate is the Jackson-Vanik amendment to the Trade Act of 1974. The amendment's original purpose was to pressure the Soviet Union to allow free emigration of Jews. Jackson-Vanik has evolved further as a weapon to fight against human rights' violations. It has been used to deny MFN to non-market economies that restrict free emigration, to China for its crackdown on pro-democracy demonstrators in 1989, and to any country condoning child labor abuses.

This linking of trade benefits to human rights' abuses has been further expanded in the minds of many in the U.S. Congress to include China's illegal sale of nuclear technology to Pakistan, its war games off the shores of Taiwan, and intellectual property piracy—arguably an interpretation far outside of Jackson-Vanik.

Those who favor delinking the issue of human rights' abuses from trade benefits argue that human rights will improve in China if trade is allowed to prosper. They further argue that MFN does not confer special U.S. tariff concessions, but only those tariff concessions the U.S. gives to almost all countries (over 160 countries), including hostile countries such as Iran, Iraq, and Libya. At this time only a few countries are excluded from U.S. MFN status, including Afghanistan, Cambodia, Cuba, Laos, Montenegro, North Korea, Serbia, and Vietnam.

Special tariff treatment, more favorable than MFN, applies to over 100 countries under such programs as NAFTA, the Generalized System of Preferences (GSP), the U.S.-Israel Free Trade Area, and the Caribbean Basin Initiative. Under these programs, designated products receive further reduced rates or zero rates.

If the U.S. withdraws MFN treatment to China, tariff rates would apply under the Tariff Act of 1930 (the highly protectionist Smoot-Hawley tariff law). Tariffs under this Act for China would increase from an average rate of 6-8% to an average rate of 40-50%. Some items would go to 100%. Tariffs on a cordless phone, for example, would increase six-fold. Tariffs on semiconductors would go from zero % to 35%.

While China MFN is controversial in the U.S., industry has consistently supported it as a way to improve U.S. relations with China. There is a certain degree of self-interest for industry in this support. If China is not granted MFN status, it could retaliate, putting the U.S. business community's multi-billions of dollars of investment at risk.

The future of China MFN is up to conjecture, but short of a major U.S.-China confrontation or extreme trade violations, it would seem that China will continue to receive yearly MFN extensions.

Understanding China's Prospects of Joining the World Trade Organization (WTO)

The foreign business community must be wary of some of the pitfalls of doing business in China. Chinese tariffs can be high, protection of intellectual property is spotty, and your sales must often be made through middlemen who take a cut of your profits. Nonetheless, there exists a great opportunity to overcome many of these obstacles through China's possible entry into the WTO.

China, with a population of 1.2 billion people, is viewed by many as the next great business frontier. Americans, Europeans and Asians are rushing in to tap this huge market. While some foreign companies are doing well in China, most are finding entry to the market to be difficult. In some product areas, entry is conditioned on setting up manufacturing operations and employing Chinese workers.

State-run enterprises are still the norm, although China is moving toward privatization of many of its industries. The Chinese government would prefer to see a strong indigenous industry, where Chinese sell locally and for export. Such local favoritism is currently not possible, however, as China lacks the capital and technology to satisfy the buying needs of its citizens. Thus, the way is open for foreign corporations.

Many in China strongly desire entry into the WTO. Foreign industries are responding to China's proposed entry. The Americans and Europeans have been especially active. These industries have talked to their trade negotiators (for example, the United States Trade Representative and the European Commission) about requiring China to agree to several conditions before it will be able to join the WTO.

Some of these conditions would require China to lower its tariffs, improve its intellectual property protection, cut out middlemen so foreign companies can sell directly and provide after sale service to their customers, adopt anti-dumping rules especially tailored to state-run enterprises, abandon its policy of forced technology transfers by foreign investors in order to do business in China, and provide foreigners with the same economic benefits as Chinese locals (the concept of national treatment) in such areas, for example, as renting office space and telephone lines.

China most likely will enter the WTO by the year 2000, or possibly sooner. Prior to its entry, you have an opportunity to aggressively weigh-in with your government trade negotiator to require China to reduce or end trade barriers to doing business in China. Employees of the United States Trade Representatives Office and the European Union are waiting for your call.

Act now!

CHINESE YUAN (RENMINBI)

HONG KONG DOLLAR

Summary

The U.S. market is open; many foreign markets are closed. Section 301 and Super 301 help open foreign markets.

Foreign dumpers, often free from competition in their closed home markets, can destroy U.S. industry. The U.S. anti-dumping laws are a tool to prevent injurious dumping.

High tariffs in foreign markets can make U.S. companies uncompetitive. U.S. industry's efforts to reduce tariffs around the world should continue as a priority.

U.S. export controls are having less effect on many companies, but we risk ceding industries producing encryption technology to overseas competitors if such controls are not eased.

Regional trading blocks, such as the European Union, will put competitive pressures on U.S. companies. NAFTA should give U.S. companies more muscle to compete with the European Union.

The China market offers many potential rewards. Foreign companies are rushing into this market. U.S. policy, especially the issue of China and MFN status, puts U.S. industry at a disadvantage with Asia and Europe. Permanent MFN status for China is one answer. A positive U.S. government policy toward China will also help.

CHAPTER TWO

COMPANY LAWSUIT EXPOSURE ISSUES

KEY TOPICS

> *Handling Lawsuit Abuse*

> *Defending Against Frivolous Securities Lawsuits*

> *Avoiding Product Liability and Punitive Damage Awards*

> *Working with the Food and Drug Administration*

> *Complying with the Clean Air Act*

> *Understanding the Clean Water Act*

> *Dealing with Leaking Underground Storage Tanks*

> *Encouraging Environmental Responsibility Through Tax Policy: A New Idea—The Carrot and Stick Approach*

COMPANY LAWSUIT EXPOSURE ISSUES

Lawsuit exposure is an important public policy area for companies as frequent litigation becomes even more common in our society. This chapter discusses the more common areas for suit—abusive lawsuits, stockholder class action frivolous lawsuits and product liability suits. Also discussed are three of the major environmental law concerns for companies: air, water, and leaking underground storage tanks, and a new environmental responsibility approach through the use of tax policy incentives.

Handling
Lawsuit Abuse

U.S. society is too litigious. Aggravating this tendency are too many unnecessary lawsuits brought to satisfy egos ("I'm going to punish you") and to leverage case settlements ("regardless of the merits of my case, they will settle just to get rid of me"). These unnecessary lawsuits are causing a drain on our courts, our society and America's competitive posture with the world.

For industry, and especially the smaller companies, it can be a deflating experience. I realize there are many meritorious lawsuits in such areas as securities fraud, product liability, theft of intellectual property and illegal activity in the environmental and antitrust areas. Nevertheless, frivolous and abusive lawsuits no doubt comprise one-half of all lawsuits today.

What can we do about this problem?

There are many facets to this problem. We can start by looking at law school admission and training standards. Part of the blame rests here. Perhaps we ought to have more admittee background checks aimed at identifying tendencies which show vindictive and game-the-system conduct. Law school teachers similarly should be judged by a higher standard.

Law schools should require more courses on ethics. Law professors should teach students to look at America's competitive well-being and weigh that against encouraging more lawsuits. Lawyers should be urged to counsel their clients to have more restraint in bringing lawsuits. Judges should decide earlier on when a case is abusive and without merit.

My experience with these abusive and frivolous lawsuits is that they are often brought by less successful lawyers trying any way they can to earn a living. Many of these lawyers have barely gotten through law school and often have been repeat bar exam takers.

While I am also a lawyer and generally proud to be in the legal profession, I am embarrassed by those in our profession who are gaming the system and essentially earning fees by "stealing" from innocent defendants through nuisance claims aimed at negotiated monetary settlements.

Local, state and federal bar associations need to crack down on lawyers abusing the system. Legislators need to enact laws to punish these abuses. One remedy would be to enact "loser pay" laws in the egregious cases. Make both the plaintiff and his attorney liable for the defendant's legal fees. If done, plaintiffs and their lawyers will make a greater effort to determine if their case has merit.

The "loser pay" concept ought to result in fewer frivolous and abusive lawsuits, thus freeing up the courts to listen more expediously to meritorious cases and let the business community do what they know best—sell their products.

Curbing these abuses is a long-term project. Industry should jump in for the long haul by supporting legislation that will stop these undesirable practices.

In addition, companies may want to stand up to frivolous and abusive lawsuits on principle. While generally a more expensive proposition than settling lawsuits out of court, this stand may be helpful in sending a message to plaintiffs and their attorneys that abusive and frivolous lawsuits will not be tolerated.

Defending Against
Frivolous Securities Lawsuits

Much of U.S. industry is subject to class action securities suits, particularly when a company's stock has had a sharp decline of 10% or more. Certain law firms use sophisticated computer programs that tell when a stock has declined, giving the firms the ability to file securities fraud or negligence actions within a few hours or days.

This quick and frequent lawsuit practice, which has hit companies in many industries, including about one-half of the high technology companies in the past few years, causes time-consuming court proceedings. The result is usually a negotiated settlement, often running in the millions of dollars.

Many corporations refer to these proceedings as frivolous securities suits brought by over-zealous lawyers who prey on industries with great cyclicality in their stock prices. While this may be true, there have been other lawsuits based on merit. Smart companies, to protect themselves, use in-house and outside law firms to lessen their lawsuit exposure in this area.

In 1995, a strong business community effort, supported by accounting and securities firms, resulted in passage of the Private Securities Litigation Reform Act of 1995. This bill, passed over the veto of President Clinton, makes it more difficult for plaintiffs' lawyers to bring such suits in federal courts.

It provides safe harbor oral and written language that can be used by corporations in their public announcements and also abolishes joint and several liability rules in favor of proportional liability rules where a defendant company's liability is limited to its level of fault.

Since 1995 there has been a decline in frivolous securities suits at the federal level. Interestingly, however, there has been an increase in these suits at the state level. Most of these suits have been in California, where the law and court decisions have favored plaintiffs and where many high tech companies are located.

The business community has become active in opposing state law and initiatives that circumvent the new federal law. The most famous initiative was California's Proposition 211, which the voters defeated in 1996 by a 4 to 1 margin after a furious business community lobbying effort (including a strong advertising blitz).

Of particular concern are proposals that limit or abolish the ability of the company or the company's insurance carrier to indemnify officers and directors for cash judgments awarded against them by a state court. Companies feel such new indemnification restrictions will impair their ability to attract qualified officers and directors.

The business community has promoted the passage of state laws along the federal model. The business community also has supported an amendment to the federal law that it preempts state law, leaving the federal law as the only applicable law in the area. Both alternatives have reasonable chances for success. Many observers predict the federal proposal—the Uniform National Standards Act—will become law in 1998.

As securities suits can be expensive and time-consuming, you may have an interest in following a typical securities suit from beginning to end.

U.S. DOLLAR

Anatomy of a High Technology Securities Lawsuit

> An electronics or biotech company announces a new product, a fore-cast of increasing sales or profits, or some other statement which encourages investors to buy the company's stock. At the time of the announcement, the stock price is $20.00 per share.

> Investors buy the stock and drive the price to $25.00 per share.

> The company then announces that the product will be late to market or that sales or profits will be below those previously forecast. The company's stock falls in one day to $20.00 per share. Investors have paper losses of between $1.00 to $5.00 per share.

> A plaintiff's securities law firm[1] brings together a small number of shareholders who join to file a class action lawsuit on behalf of all "injured" shareholders against the company in federal court. The law firm files its complaint against officers, directors, and key company employees alleging securities fraud under Section 10b and Rule 10b-5 and insider trading violations under Section 20A of the Securities and Exchange Act of 1934. Also commonly alleged are fraud, deceit, and negligent misrepresentation violations under local state law.

> The company, depending on the merits of its case, its cash resources, and the amount of time its management staff has to answer legal plead-ings, including requests for company documents, interrogatories, and depositions, will either carry the case to an actual court judgment or settle before a judgment without admitting fault. An incentive to settle is that directors' and officers' insurance policies (D&O insurance) generally will not indemnify a finding of liability (and cash judgment) against an officer or director found to be engaging in fraud under Section 10 of the Securities and Exchange Act of 1934. Most compa-nies agree to settle in such situations rather than subject their officers and directors to the risk of no indemnification and consequent poten-tial loss of their personal assets.

---

[1] Plaintiff securities law firms can file their class action lawsuits quickly because they hire professional plaintiffs who buy 1-2 shares in each of the companies listed on the major stock exchanges. These firms also utilize sophisticated programs that can track company announcements and 5-10% sudden declines in company stock prices. Because of the professional plaintiffs and the computer tracking systems, a law firm can often file a lawsuit in 1-3 days.

> The court hires a special securities litigation processing firm to process all claims and pay cash awards.

> The injured shareholders typically receive 10% to 60% of the paper loss claims they filed with the court. This recovery may be different from the shareholders' actual gains or losses. For example, if the shareholders sell one year later for $30.00 per share, they will have a $5.00 to $10.00 gain over the $20.00 per share price alleged to cause the shareholders' injury. On the other hand, if the shareholders sell one year later for $10.00 per share, they will have an actual loss of $10.00 per share relating to the $20.00 per share injury price plus a loss of $1.00 to $5.00 per share relating to their actual purchase price over $20.00 per share.

> The plaintiff law firm typically receives one-third of the settlement or court judgment. The securities processing firm and others working on the case receive payment of 3 to 5 percent of the settlement or court judgment.

> The time elapsing from the settlement or judgment to the pay out to injured shareholders is generally 12 to 24 months.

Avoiding Product Liability and Punitive Damage Awards

A combination of far-reaching laws, meritless suits, and high punitive damage awards have made the introduction and sale of your products a sometimes risky business. There are a plethora of different product liability laws in the fifty U.S. states and territories. Some carry the potential of full liability for an injury, even when your company is less than 1% at fault. These issues, plus the high legal fees and countless other hindrances to business that lawsuits present, have led the business community to seek ways to make product liability laws simpler and fairer.

The platform advanced by the business community consists of the following major areas:

> **Standardizing Product Liability Laws.** At the federal and state levels, the product liability laws are very different. Standardizing them would save companies legal costs and reduce confusion in interpreting the different laws.

> **Ending Abusive Punitive Damage Awards.** Punitive damage awards have increased greatly in recent years. These are awards that the courts may impose when actual (compensatory) damages are deemed inadequate. The business community platform would limit punitive damage awards to cases involving, for example, reckless behavior or knowingly bad conduct, in situations where there is "clear and convincing evidence" of such behavior. If such bad behavior is found, the award could be limited to 2 or 3 times the actual damages, or $250,000, whichever is greater.

> **Apportioning Damage Awards.** Abandon the "deep pocket" doctrine which can impose liability on multiple defendants out of proportion to their level of fault. For example, if a computer company sells a computer to a state agency, and an employee of that agency is hurt when the computer overheats and blows up, it is possible the parties sued would be the computer company and the state agency. If the computer company had faulty connectors and was 95% at fault while the state agency was 5% at fault for having a poor electrical power outlet, and if the computer company has gone into bankruptcy unable to pay its damage award, the state agency theoretically could be required to pay 100% of the damages because of its cash resources (deep pockets). An apportioned liability system would end this "joint or several liability" method and attribute damages in proportion to the level of fault. Thus, in this case, the state agency would be only 5% responsible for damages. To date, 38 states have enacted "fair share" liability laws.

> **Ending Product Liability Suits for Old Property.** If equipment purchased 15 or 20 years previously becomes defective, a statute of limitations provision would come into play eliminating a product liability award after this period of years.

> **Shortening Statute of Limitations Period for Filing of Claims.** Require plaintiffs to file suit within two years after they discover, or should have discovered, the harm and its cause.

> **Having Losing Party Pay where a Product Liability Suit is Determined by a Court to be Frivolous, Without Merit, and a Sham.** In those situations where a defendant is unjustly sued in a meritless action, without any factual basis, as a way to extort a negotiated damage award settlement against the defendant, the court could ask the plaintiff to pay the defendant's legal fees.

There has been legislation at the federal and state levels to adopt this business platform. Public sentiment is beginning to favor the business community. The trial lawyers and unions generally have opposed reform. President Clinton has stated on several occasions he will use his veto power to oppose reform.

Companies should act now to be part of the product liability solution and help build momentum in the U.S. Congress to override a presidential veto.

Working with the Food and Drug Administration

Perhaps the most important public policy issue for the biotechnology and the medical equipment industries is the length of time it takes to bring new products to the market. The time from product development to market introduction can take as long as fifteen years.

Lengthy lead times are a requirement of the Food and Drug Administration (FDA), whose role is to insure that the public does not use dangerous drugs. Indeed, the biotechnology and medical equipment industries appreciate a conservative approach considering the potential for product liability exposure down the road. Recent deaths allegedly resulting from the use of diet and anti-impotence pills illustrate this point.

Testing products for so many years can be expensive. Why does product testing take so long? A look at the biotechnology sector is useful as an example.

There are several major phases of product development before a drug can be sold to the public.

> Research phase — identifying the chemical composition to treat a disease.

> Development phase — working on and improving the chemical composition, manufacturing it into a drug, and testing it on animals.

> FDA approval phase — getting FDA go-ahead to test the drug on humans.

> Phase 1 — analyzing the safety of the drug.

> Phase 2 — determining the drug's efficiency and effectiveness.

> Phase 3 — proving that the drug works in controlled human trials where some take the drug and some take a harmless sugar pill (placebo).

> New drug application process — presenting to the FDA the results of the data collected.

These development phases may cost the drug company hundreds of millions of dollars, take an average of 8-15 years and require the company to put together over 100,000 pages of documentation for FDA review.

At Table 2.1, I consider a standard type of drug and illustrate at various product development phases a typical time range necessary for each phase and a typical patient sample size for the human testing phases.

Table 2.1 Product Development Chart

PRODUCT DEVELOPMENT PHASE	LENGTH OF TESTING PERIOD	PATIENT SAMPLE SIZE
Research	1 - 3 years	
Development	1 - 2 years	
FDA Approval to Test on Humans	3 months to 1 year	
Phase 1	6 months to 1 year	60 - 120 people
Phase 2	1 - 2 years	120 - 140 people
Phase 3	2 - 3 years	1,000 - 10,000* people
New Drug Application Phase	2 - 3 years	

* sometimes up to 40,000 people at a cost to the drug company of $2,000 to $5,000 per person

After the product is finally available on the market, drug companies face still more obstacles. Will the government permit reimbursement to the patient of the cost of the drug for Medicare purposes? Will the health insurers permit reimbursement?

The issue often comes down to an analysis of whether the drug is life saving versus quality of life enhancing, as, for example, hair growth or anti-impotency drugs. Economics plays a critical role. The government or a health insurer might determine it cannot afford to reimburse patients for expensive quality of life drugs that could potentially be used by a large segment of the population.

Both the biotechnology industry and the FDA are trying to accelerate all phases of the FDA review process period. Congress is also getting involved as the public seeks faster-to-market solutions in such areas as AID's, cancer, heart, Parkinson's and Alzheimer's research. The goal is to get the product to the market as fast as possible with the least product liability risk.

Complying with the Clean Air Act

The Clean Air Act sets national air quality standards and attainment dates for state compliance. Any changes in the standards may require greater pollution control expenditures by business.

Alert management uses in-house and outside environmental law specialists to help insure that their company is in compliance with the environmental laws. Air pollution varies from state to state and from region to region within states. High technology companies in Silicon Valley, for example, have complained that they have been subject to the stringent compliance standards set for the smoggier Los Angeles area even though the San Francisco Bay Area/Silicon Valley air quality is typically among the best in the country for a metropolitan region.

The business community, a strong supporter of cost-effective approaches to cleaning the air, has been pushing for more selective compliance standards to meet clean air guidelines. Why should business, it is argued, spend billions of dollars on pollution control facilities when the answer, in great part, could be the removal from the highways of pre-1980 cars known for lesser pollution compliance standards?

The business community has been cooperating with government agencies to enhance air quality. In one situation, companies put money into a pool to purchase and permanently scrap the most air polluting cars. In another case, companies supported the use of cleaner burning gasoline and further reductions in tailpipe emissions. And, according to various studies, many industries have substantially reduced air pollutant emissions from their manufacturing sites.

The business community will continue to seek ways to ensure clean air. To be cost-effective in meeting these standards, however, will require less government bureaucracy and consistent application of similarly worded federal, state, and local regulations. Those localities having the least amount of bureaucracy will be favored by companies trying to identify new manufacturing sites.

Understanding the Clean Water Act

Industry uses (and treats and discharges) much water in its manufacturing. Virtually everyone—industry, the public, environmentalists, and local, state and federal governments—is concerned about the impact of manufacturing chemicals and processes on the environment, particularly water resources.

The Clean Water Act addresses many of these concerns. Another federal law, the Safe Drinking Water Act, addresses minimum quality standards of drinking water. The issue of leaking underground storage tanks and their potential to contaminate groundwater is another important issue and is covered in the next topic.

There are two critical public policy issues regarding water in manufacturing. The first is the overlapping jurisdiction for regulation of water supply, cleanup, and industrial waste water discharge by federal, state, and local authorities. The second is the question of "How clean is clean?" when addressing treatment of contaminated ground or surface water and pretreatment of industrial waste water.

The business community advocates more efficient government protection of water resources. Overlapping, and sometimes conflicting, water-related laws and regulations make compliance complicated and costly. Consolidating regulatory oversight committees and standardizing compliance requirements will improve protection of the environment and public health and safety while it saves industry time and money.

The issue of determining appropriate cleanup standards for contaminated groundwater or surface water is a challenging one. Is it feasible to restore the water supply to its "original" purity? Or should it be restored to drinking water standards which may be more or less pure than its original state? Or are other standards appropriate?

A variety of considerations specific to each individual cleanup activity may make any one of these options the best choice. Parameters to consider include the geology of the site, the nature of the contamination, the past, present and probable future uses of the water body and the site, the technology available to address the contamination, the cost to implement each technology at the site, and the likelihood that the selected standards can be achieved in a reasonable time frame. With the encouragement of industry, government agencies are tending to weigh more carefully the risk to public health and the environment when they choose a cleanup standard.

Dealing with Leaking
Underground Storage Tanks

The discovery of groundwater contamination in Silicon Valley in the early 1980s caused the high technology industry's "clean" reputation to be challenged. Industry, government, and the community all questioned the practice of storing such potentially dangerous materials as gasoline, diesel fuel, home heating oil, and industrial solvents in underground tanks. A nationwide investigation was undertaken to determine the extent to which underground water resources have been impacted by this practice, which had been considered environmentally benign.

Silicon Valley industry has actively participated in a coalition with public officials and community representatives to produce a model Hazardous Materials Management Ordinance (HMMO). Its purpose was to prevent future leaks based on the new information gained through the investigation of contaminated groundwater sites. Adopted in 1983, the HMMO served as a model for state-wide requirements put in place later in that year. The subsequent federal underground storage tank program, also based on this model, requires every tank in the country to be upgraded to a double-walled type by 1998.

Since the early 1980s, companies and individuals have spent billions of dollars to clean up groundwater contamination and to prevent it from impacting drinking water wells. In California alone, the costs have exceeded one billion dollars. In some cases, the conservative cleanup rules have bankrupted independent gas stations, resulting in a one-third decline in the number of such stations.

Regardless of the chemical released, the cleanup of a leaking tank typically involves the removal of the tank and any surrounding soils contaminated by the leak. Soils must be treated until "clean," as determined by the oversight agency. Groundwater must be tested to determine whether it has been impacted. Contaminated groundwater is typically pumped out of the ground, treated until it is "clean" (as determined by the oversight agency), and ultimately disposed. These remediation activities can often last years and cost millions of dollars. In large part, the length of the project and the magnitude of the investment are determined by the cleanup standards, as discussed in the preceding topic.

A 1996 study in California by the Lawrence Livermore National Laboratory, funded primarily by the U.S. Environmental Protection Agency (EPA), now casts doubt on whether the billions of dollars spent to clean up leaking underground fuel tanks were necessary or worthwhile. Some of the major findings include:

> Gasoline from leaking tanks rarely spreads more than 250 feet underground, much less than previously assumed.

> Microbes naturally occurring in the soil often can consume gasoline as quickly as pumping.

> Of over 12,000 public drinking water wells examined in the study, only six had been found to be contaminated by leaking underground fuel tanks.

State regulators are already reviewing the current ground water cleanup requirements to determine where more practical, yet protective approaches might be implemented. Industry's support of this move toward evaluation of the relative risk to individuals versus the cost-effectiveness of the treatment is based on a decade of evidence that public health and the environment are not adversely impacted by underground storage tanks.

Encouraging Environmental Responsibility through Tax Policy:
A New Idea—The Carrot and Stick Approach

Scientists are telling us that emissions of greenhouse gases are changing the world's climate. If this trend is not reversed, global warming will adversely impact our planet.

Greenhouse gases include carbon dioxide, ozone, methane, nitrous oxide, and Freon. Carbon dioxide generated by burning oil, gas and coal make up 50% of all greenhouse gases. Transportation vehicles make up 50% of the carbon dioxide emissions.

What is the danger of global warming? Arguably because of global warming the majority of the world's glaciers are shrinking, leading to higher sea levels and newly flooded land masses, new diseases at higher altitudes and latitudes, and perhaps longer disruptive El Niños than in past years. The economic burden on society because of the flooding and new diseases will be staggering in the amount of property damage and loss of lives.

How can we encourage people to take the environmental consequences of their actions into account? How can we induce business to concentrate on the adverse effects of chemicals on the ozone layer and carbon dioxide and other greenhouse gases on global temperatures?

There are no simple answers to these questions. Reasonable people will differ as to the threat, urgency and solutions necessary to deal with the problems arising from these chemicals.

One approach is gaining momentum. It is to change people's environmental habits through tax policy. It is the carrot and stick approach.

First the stick. Government should collect new tax revenues through a variety of pollution taxes. Tax ozone-depleting chemicals. Tax carbon dioxide and other greenhouse gases.

So where is the carrot? The carrot could be a reduction of payroll taxes and individual and corporate income tax rates.

The concept here is that market forces should prevail to price energy products in an amount that includes their air pollution damage and climate change costs. Pollution taxes on top of normal product costs will achieve the goal of a true market price for these products.

At the same time, lower taxes should apply to individuals and businesses. *The overall goal is a revenue-neutral tax shift, not a tax increase.* Thus, to the extent energy taxes increase, individual and business taxes should decrease proportionally.

Many feel that properly pricing energy through energy tax reform, in combination with fundamental tax reform, will create benefits for the economy. Leading the way are a number of European countries. Through numerous energy taxes, they have reduced the use of pollutants and at the same time payroll taxes. American business is giving a serious look at the carrot and stick approach. Among industry leaders are Hewlett-Packard and many in the high tech field.

Over the years this carrot and stick trend will continue and probably accelerate. It is time for the business community to actively step up and be part of the solution.

FRENCH
FRANC

GUATEMALAN
CENTAVO

MEXICAN
PESO

EL SALVADORIAN
CENTAVO

Summary

The threat of abusive and frivolous lawsuits is helping to bond the business community like never before. While a law helpful to the industry was passed at the federal level in 1995, securities trial lawyers are attempting to modify state laws as a way to override federal law.

Product liability will continue to be a major public policy issue. Industry should work to change product liability laws so that they are less burdensome and costly.

Environmental issues appear not to be the threat to industry that they were in the 1980s as a result of the great lengths to which industry has gone to equip its factories with the latest air and water pollution control equipment. The major issue today is the conflicting maze of federal, state, and local environmental laws that add greatly to industry costs and burdens. An emerging new approach —encouraging environmental responsibility through tax incentives—appears to be the wave of the future.

MEXICAN PESO

C H A P T E R   T H R E E

ANTITRUST AND INTELLECTUAL PROPERTY ISSUES

KEY TOPICS

> *Antitrust*

Utilizing the Antitrust Laws for Research and Development and Production Joint Ventures with Your Competitors

Busting Monopolies: Making Competition Fair

Working with the Antitrust Laws of the U.S. and European Union: A Comparison

Taking Advantage of Telecommunications Reform

> *Intellectual Property*

Coping with Submarine Patents

Keeping up with the Internet and the Copyright Laws

Aggressively Using the Law to Fight Intellectual Property Pirates

Recognizing Other Types of Piracy

Stopping the Misappropriation of Trade Secrets

Protecting Semiconductor Mask Works from Piracy: The Semiconductor Chip Protection Act

Collaborating with the U.S. Government to Protect Your Intellectual Property: "Special 301"

Using the Intellectual Property Statutory Protection Periods

ANTITRUST AND INTELLECTUAL PROPERTY ISSUES

As companies grow, they must be aware of antitrust issues (where they can sue or be sued) and intellectual property issues, which can be their lifeblood. This chapter discusses some of the more recent public policy issues in these areas.

Antitrust

Utilizing the Antitrust Laws for Research and Development and Production Joint Ventures with Your Competitors

The U.S. antitrust law historically has caused companies to be hesitant about forming joint ventures with competitors. To be accused of controlling production or engaging in price fixing or allocating markets, customers, or products—or even of monopolizing an industry—could bring criminal penalties and civil damage awards of triple times the injury.

As a result, the concept of an exception for research and development ventures among competitors became more attractive. Most major foreign competitors already permitted such activities. Research and development was becoming expensive; it was wasteful for several companies to each be doing the same expensive research and development activities on their own. In fact, some projects were so expensive they could not otherwise be undertaken without a joint venture.

It was in this context that industry worked to pass legislation in 1984 to permit research and development ventures among competitors. This legislation is known as The National Cooperative Research Act of 1984. Joint ventures benefiting from this legislation include the Microelectronics and Computer Technology Corporation (MCC) and SEMATECH, a two consortia of high tech companies working on the latest semiconductor manufacturing techniques.

There have been over 100 joint ventures using this law since 1984. Some of the major specifics of the law are as follows:

> Pro-competitive joint ventures will be favored.

> If a court finds anti-competitive aspects of the joint venture to exist, triple damages will usually not apply. Single (actual) damages would be enforced.

> Most research and development joint ventures will be allowed, provided the joint venture will not be in a monopoly position and legal notice rules under the law are followed.

> The plaintiff (the one charging the antitrust violation) can be held liable for the defendant's legal fees if the case was brought under circumstances showing a frivolous or sham purpose.

In 1993, a successor law was passed, this time to permit production joint ventures among competitors. The new law, The National Cooperative Production Amendments of 1993, provided many of the same benefits as the 1984 law. It was passed in response to a perceived lack of manufacturing capability of U.S. industry and the fact that building state-of-the-art manufacturing facilities was becoming very expensive. By the year 2000, for example, many semiconductor wafer fabrication facilities will cost two to three billion dollars to build, making it absolutely critical to be able to do production joint ventures for this sector.

This is a public policy area that needs to be watched. The 1984 and 1993 laws were passed over objection of many in the antitrust bar. If, in the future, these ventures appear to be leading to anti-competitive behavior, they could be subject to attack. Be vigilant.

CANADIAN DOLLAR

Busting Monopolies: Making Competition Fair

Major U.S. antitrust legislation has been in place for over one hundred years. The two most important laws are the Sherman Act of 1890 and the Clayton Act of 1914. They were enacted to stop monopolistic, uncompetitive behavior principally of the oil, railroad and steel companies.

These companies, to insure little or no competition, used various tactics to drive competitors out of their markets. Industries were concentrated in the hands of a few companies, prices were fixed to prevent competitive underbidding, and markets and product lines were divided to insure freedom from competition.

When there was competition, you could beat out your competitor through predatory pricing, i.e., pricing below cost until your competitor, without your financial staying power, could no longer absorb the losses and had to exit the market. Another way was through tying arrangements, whereby a dominant oil company could, for example, refuse to sell you gas and oil unless you bought its tires, batteries and accessories also.

The Sherman and Clayton Acts were successful in stopping many of the above uncompetitive practices and still remain the main antitrust enforcement tools today. When antitrust violations are found, penalties can sometimes be as high as three times the actual harm incurred.

Some would say antitrust enforcement is an abuse of government power. Let capitalism and the markets work. However, for a small- or medium-sized company, antitrust enforcement can sometimes mean the difference between going out of business and survival. These smaller companies generally do not have the ability to fend off a bigger competitor's predatory pricing or its tying arrangements.

While the public has witnessed tough antitrust enforcement in this century, there have been periods of laxness. Such was the case during the President Reagan era of the 1980s and the early 1990s. During that time many mergers and acquisitions passed U.S. government (U.S. Department of Justice and Federal Trade Commission) scrutiny. This was a period of merger and acquisition activity often involving companies in different industries. And when there were new combinations, the remaining competitors typically still numbered in excess of five-to-ten companies. Antitrust enforcers saw this type of activity as not harmful to consumers.

Over the past few years, however, the environment has changed. Recently we have seen mergers, acquisitions and "strategic combinations" of companies in the same industry. These latter virtual companies have the benefit of new market mass without the difficulties of melding company cultures, incurring extra taxation, and impacting financial agreements and stock prices. The airline industry represents a good example of these strategic combinations.

Where there used to be five-to-ten competitors in an industry, now sometimes there are only three-to-five. One has only to look at recent activity in the airline, defense and accounting industries to document this point.

While many combinations still are being approved, others are finding the going a little rough. Recent major combinations that have been approved include Bell Atlantic and Nynex, McDonnell Douglas and Boeing, Southern Pacific and Union Pacific, Chase Manhattan and Chemical Bank, and Price Waterhouse and Coopers.

Other combinations are, however, undergoing rigorous review. These currently include Lockheed Martin and Northrop Grumman, WorldCom and MCI, Northwest and Continental, Compaq and Digital Equipment Corporation, BankAmerica and NationsBank, Daimler-Benz and Chrysler, and SBC Communications and Ameritech. Strategic combinations, such as British Airways and American Airlines, also are having trouble. And Ernst & Young and KPMG Peat Marwick called off their merger because of government opposition, this time from the enforcement division of the European Union.

Along with this recent tougher government scrutiny of

mergers and other combinations, the government has also attacked Microsoft for illegal tying of its Internet browser to its market-dominant Windows operating system, for predatory pricing of its browser, and for other monopolistic practices designed to eliminate potential competitors.

The government is also looking closely at Intel's dominance of the semiconductor market, especially allegations it threatened to cut off smaller competitors or customers with which it had legal disputes from its top-of-the-line chips.

The good news for small- and medium-sized companies is that they have the government in their corner more often today. The bad news is that it is expensive and very time-consuming to bring an antitrust case. Zenith Corporation is a case in point. They charged Panasonic Corporation with antitrust abuses in the TV industry in the 1970s. The case lasted about eighteen years, Zenith did not win, and during this time period Zenith went from a TV maker powerhouse to an also ran.

While you may feel good that the government is behind you, if you put the shoe on your other foot, you may not enjoy the thought that someone can more easily sue you today for antitrust violations.

Will this increased antitrust activity help society? One could seriously ask the question—did the breakup of the Bell operating companies in the 1980s increase competition and efficiency for telephone service, or did airline deregulation in the 1970s increase competition, reduce fares and improve quality of service in the airline industry?

Whether the latest round of stronger antitrust enforcement will increase competition and benefit consumers with lower prices, more product innovations and better service may be answered in the next decade.

The moral of this story for you and your company is clear. You must now exercise more prudence not to violate the antitrust laws (U.S., EU and perhaps Japan). You need competent antitrust lawyers to advise you. Prevention is all important. Control your recklessness.

Working with the Antitrust Laws of the U.S.
and European Union: A Comparison

U.S. antitrust policy stems in great part from the Sherman
Act of 1890, which consists of Section 1, dealing with
restraints of trade, and Section 2, which deals with monopolies. European Union (EU) antitrust policy (called Competition
Policy) is dealt with primarily in Articles 85 and 86. Article 85
prohibits restraints of trade while Article 86 prohibits abuse
of dominate position. The U.S. Department of Justice and the
Federal Trade Commission enforce U.S. antitrust policy; the
European Commission (EC) enforces EU antitrust policy.

U.S. and EU antitrust policies generally prohibit price fixing
for purchases and sales of products, limiting or controlling production, allocating market territories, products, or customers,
and the entering into of "tying arrangements." An illegal tying
arrangement could be the result of a situation where a dominant memory chip vendor conditioned (tied) its sales of memory chips to those buyers who agreed to buy a certain amount
of microprocessor chips as well.

Article 86 appears similar to Section 2 of the Sherman Act
in attacking possible monopoly situations where a company
abuses a dominant market position in a relevant market
and product area. Article 86 may go beyond Section 2 of
the Sherman Act by treating predatory pricing (pricing below
total cost as well as below variable cost) by a dominant competitor as illegal since predatory pricing could drive companies
with equal or better products, but with less financial staying
power, out of the market. The U.S. appears not to use the
predatory pricing concept to a major extent.

Both U.S. and EU antitrust policies apply to transactions
either within or outside their respective borders which impact
competition within their borders. This extraterritorial reach
means, for example, that U.S. price-fixers of products entering
the EU could be in violation of U.S. law and EU law.

When planning a European transaction, it is often prudent

to review U.S. law, EU competition law, and laws of the EU member states not superseded by EU competition law.

Violation of U.S., or EU law, or laws of the EU member states may involve penalties. The U.S. practice of awarding damages equal to three times the injury ("treble damages") appears more extreme than the fines imposed on violators by EU and member states' laws. U.S., EU, and member states' laws all have provisions to stop the illegal activity. The U.S., however, also provides in its law the imposition of criminal penalties, including jail terms. The EU and member states do not impose criminal penalties.

If antitrust violations are a potential issue, a company may ask for an exemption when the impact on trade and competition is minor. In the U.S., a company may ask for an exemption through the "business review procedure" of the U.S. Justice Department or the Federal Trade Commission. In the EU, the exemption may come from "block exemptions" applying to certain general activity areas and from negative clearances and comfort letters applying at the individual transaction level.

Awareness of the law and caution in this area are important. Knowing how to work the political aspects within the U.S. Congress and Administration and the EC in the EU may be equally important here.

Taking Advantage of Telecommunications Reform

In 1996, Congress wrote the first major reform of the nation's telecommunications policy in over sixty years. By an overwhelming vote in the House of Representatives and the U.S. Senate, Congress passed the Telecommunications Act of 1996, an attempt to increase competition while protecting the public.

The Act encourages more competition among the major telephone and cable companies, as well as encouraging new entrants at the local level, by phasing out government-supported monopolies. Open competition will provide consumers with more choice of services in the continuing meshing of the telecommunications, video, and computer industries.

With the signing of the Act, regional phone companies, such as Pacific Telesis and Bell Atlantic ("Baby Bells"), will be permitted to enter the long-distance market. Long-distance companies will be permitted to enter the local telephone market. Cable TV companies may offer local telephone service. Telephone companies may provide video services. Mergers between local, long-distance, and cable companies are now possible. Already, mergers have taken place between Pacific Telesis and SBC Communications and between Bell Atlantic and Nynex. Other merger activity will take place, including the recently announced merger of SBC Communications and Ameritech.

To avoid unnecessary upheaval to these industries and consumers, the Act provides transition periods for their deregulation. While proponents praise the deregulation attempts of the Act, citing such long-term benefits as increased consumer choice, lower consumer prices, more efficiency in the industry, and greater technological and investment potentials, detractors point to the great number of new rules the Federal Communications Commission (FCC) will need to adopt in order to comply with the Act's provisions.

Up for debate are provisions calling for the rating of TV programs and the placing of V-chips inside TVs to permit the screening out of objectionable material such as violence and pornography.

The American Civil Liberties Union (ACLU) objects to the provisions of the Act which limit the flow of information over the Internet. While one U.S. court has declared this provision unconstitutional, this will continue to be a hot issue. Consumer groups object to parts of the Act which, they claim, will result in mega-mergers and fewer competitors, thus leading to increased prices and less efficient services that the U.S. anti-monopoly statutes were designed to prevent.

After two years since enactment of the Act, is deregulation

lowering prices to users? The results are mixed. Residential users complain their rates have increased, due in part to less competition among local telephone companies. Some businesses report their rates have decreased due to more competition for their business.

Over the next few years, as telecommunications companies and consumers try to adapt to deregulation, many companies will find themselves in the midst of turmoil. Some will profit and some will lose as they try to tailor their product portfolios to the winners.

Intellectual Property

Coping with Submarine Patents

Submarine patents are patents deliberately delayed by the applicant in order to extend patent expiration dates. Two famous examples involved inventors Jerome Lemelson and Gilbert Hyatt.

Lemelson claimed to hold numerous patents on bar code and robotics technologies. But one of his patent applications, filed in 1954, was not actually granted until 1994. Meanwhile, Lemelson collected about $500 million in royalties in the late 1980s and early 1990s.

Hyatt claimed to have invented the computer-on-a-chip in 1970, one year prior to a similar claim by Texas Instruments. While no one in the semiconductor industry had heard of Hyatt's claim until 1990, he reportedly collected $70 million in royalties between 1990 and 1992.

The issue of submarine patents was addressed in the recent Uruguay GATT Round, when GATT members adopted a rule providing inventors patent protection for 20 years from the date they filed their patent application. Countries that had been giving patent protection from the date of grant conformed to the new 20-years-from-date-of-filing law. (Prior U.S. law provided for patent protection of 17 years from the date of grant.)

Many inventors, particularly from the biotech sector, objected to the 20-years-from-date-of-filing provision. Biotech inventors claimed that their applications were often subjected to five- to ten-year delays, thus cutting their protection by up to half the period. Other inventors claimed the 20-years-from-date-of-filing provision would not protect them against foreign competitors and unscrupulous pirates.

Arguments for the two sides break down as follows:

Proponents of date-of-grant	a) a 20-year patent term from the date of filing is not extendible, thus causing a shorter protection period for delayed patent grants.
	b) independent inventors and small businesses will be harmed. By keeping patent applications quiet and delaying their grant, the inventor can prevent others from knowing about, and perhaps copying, the invention.
Proponents of date-of-filing	a) cures submarine patent problem.
	b) conforms to worldwide date-of-filing standard.

Both sides have valid arguments. The goal is to balance the individual and public interests. A possible solution to help the proponents of date-of-grant would be to allow reasonable extensions of time in cases where the patent office causes a delay. This approach would presumably satisfy the date-of-filing proponents.

Another compromise approach would protect inventors from foreign competition and copycats. In this approach, U.S. law would conform to more favorable foreign law which requires public notification of the patent application after 18 months.

Currently, U.S. inventors who file for patent protection overseas have their applications published in the foreign country within 18 months, while no notification is required in the U.S. The early publication requirement for applications in foreign countries allows foreign inventors to copy an invention before its U.S. inventor has been granted the patent in the U.S. If there were a similar 18-month notification in the U.S., other inventors would be put on notice and could either determine an alternative to the claimed technology or seek to license it at a time when the value of the patent is not yet clearly established and royalty rates may be more reasonable.

Many consider the submarine patent merely a matter of using the patent laws to the advantage of the patent owner. Others think of it as an unfair windfall to the owner. Whatever your view, the issue of submarine patents will find continued discussion in the literature for several years to come.

Keeping Up with the Internet and the Copyright Laws

With the rising popularity of the Internet has come a host of unanswered issues about the intellectual property rights of users and providers. One can imagine potential patent and trademark issues, but the major area of concern is the application of the U.S. copyright laws.

It is apparent that we will need to modify the copyright law to deal with the Internet. Some questions that will need to be addressed in the next several years might well include the following:

> If I want to put this book on the Internet, what kind of protection will I have against the person who wants to copy it without paying me a fee?

> If someone copies my book, what protection do I have that the copier will not sell the copies to readers around the world?

> If a teacher at the University of Texas wants to copy my book, will the doctrine of "fair use" permit him to distribute part or all of my book to his students? (The doctrine of "fair use" has permitted a limited copyright exemption for educational institutions to copy a work without paying royalties to the author.)

> While I may have some U.S. copyright protection for my book on the Internet, what about copyright protection in Uzbekistan when someone there downloads my work from the Internet?

> What is the copyright liability, if any, for an on-line provider, such as America Online, if it publishes something that unknowingly violates the copyright law? Will the provider be able to argue that the situation is

no different from a telephone company's freedom from responsibility for what goes over the telephone lines? Or is it more like a network broadcaster, such as NBC or CBS, which may be liable for its content?

> What about access-only providers to the Internet? Should permission for access to the Internet create a copyright liability for this provider?

I have presented some of the copyright issues relating to an author of a book. The copyright issues also extend to many others, including software creators and the entertainment industry. Finding answers will be difficult and time-consuming. New laws and regulations will want to strike a balance between protecting rights of authors and permitting the Internet to live up to its potential. It's an issue that involves not only the U.S., but the whole world. The U.S. will need to enter into treaties with other countries to spell out the copyright protection here. Eventually, the World Trade Organization (WTO—formerly GATT) will need to address these issues.

Aggressively Using the Law to Fight Intellectual Property Pirates

A U.S. intellectual property provision aiming at piracy is found in Section 337 of the Tariff Act of 1930. Section 337 typically has been used by U.S. companies to protect their intellectual property rights from infringement by foreign manufacturers. Under the law, Section 337 can prevent infringing goods from entering the U.S. market, making it an important defense against foreign competitors.

Infringement can take many forms, including copying another's design and continuing to use intellectual property after a license to use it has expired.

In one mid-1980s case, Texas Instruments sued eight Japanese companies and one Korean company for infringement of DRAM memory chips when the companies continued to use Texas Instrument's DRAM technology to manufacture DRAMs after the Texas Instruments' license agreement had expired. What's more, the infringing Japanese and Korean

DRAMs were entering the U.S. market. Texas Instruments filed its Section 337 case with the International Trade Commission (ITC) to stop the entry into the U.S. As a result, the Japanese and Korean companies eventually settled by entering into new license agreements favorable to Texas Instruments.

Section 337 has not been without its problems, however. A 1988 GATT panel ruling found Section 337 inconsistent with the GATT and ordered a change. The GATT panel found foreign products entering the U.S. that were alleged to infringe U.S. patents were treated less favorably in U.S. federal district court proceedings than were infringing products of U.S. origin. This difference in treatment was found to violate the GATT's equal treatment provisions.

In 1994, the U.S. passed revisions to Section 337 that will make it GATT/WTO compatible. From a U.S. public policy standpoint, it is desirable to keep in the U.S. law a Section 337 to prevent goods produced from pirated intellectual property from entering the U.S.

Recognizing Other Types of Piracy

Piracy of intellectual property is one of the most important public policy issues for the business community. Knock-offs of semiconductor designs, computer technology, software programs, pharmaceutical products, chemical processes and other technologies are costing the U.S. industry billions of dollars in lost sales and profits. A listing of the most serious country violators is shown in the topic paper under "Special 301."

There are also many cases where intellectual property protection cannot be enforced. If the pirate does not sell in the U.S., Section 337 will not apply. If the pirate's home country does not have intellectual property protection or enforcement in such areas as patents, copyrights, or semiconductor chip designs, the pirate is free to roam.

Sometimes U.S. manufacturers must use non-traditional methods to protect their intellectual property. In one recent case, a foreign competitor of a U.S. semiconductor company copied the U.S. company's chip design in a country without a chip protection act. The foreign competitor did not sell in the U.S.; thus, Section 337 did not apply. It sold only in its home country and in countries without intellectual property protection. The remedy? Moral persuasion. The U.S. company went to the U.S. Embassy in the foreign country and asked it to talk to the foreign competitor. The U.S. embassy also talked to the foreign government and to trade associations in which the foreign competitor was an active and "upstanding" member. The U.S. company was prepared to play out the piracy issue in the press (the major business newspapers and business journals). Under this pressure, the foreign competitor withdrew its pirated product from the marketplace. This was an unusual example of moral persuasion.

In another case, a U.S. semiconductor company obtained a permanent injunction against a U.S. manufacturing site of a foreign competitor who had hired key marketing and technical employees from the U.S. semiconductor company to produce an exact copy of a memory chip product. The U.S. semiconductor company brought a misappropriation of trade secrets case and got the injunction forbidding the foreign company to use these employees for designing a product competitive with the U.S. company's memory product. More information on misappropriation of trade secrets follows in the next topic.

Undoubtedly, methods of piracy will become ever more ingenious and high technology companies should be vigilant. Be prepared to use the U.S. legal system to protect your rights. And where protection is inadequate, use moral persuasion or appeal to the U.S. Congress and Administration to stop piracy through new laws.

Stopping the Misappropriation of Trade Secrets

Misappropriation of trade secrets of companies is viewed as a serious problem. Whether the misappropriation comes in wrongfully obtained copies, sketches, photographs, downloads, uploads, or employee knowledge, the cost to the U.S. community is estimated to be in the billions of dollars. The misappropriators may include a U.S. company's employees, its former employees, a foreign government or its agents, and foreign companies or their employees. There have been some important developments regarding the policing of trade secrets.

You were informed earlier in this chapter about the recent California case involving a U.S. semiconductor company obtaining an injunction against a U.S. subsidiary of a foreign company for hiring away key product line employees of the U.S. company. There have been several other cases involving hiring a company's employees to gain access to that company's trade secrets.

Recent proposals at the state level tighten trade secrets law and, in some cases, expand criminal penalties against misappropriation. The California legislature recently passed a proposal to increase the criminal penalties for misappropriation of trade secrets. In Texas, the business community successfully lobbied to have the state conform more closely to the U.S. Model Uniform Trade Secrets Act.

A major issue in Texas had been a short statute of limitations period for discovering and bringing a trade secrets case. In most other states, the two-year statute of limitations commences to run on the date the owner discovers his trade secret has been misappropriated; Texas commenced its two-year statute within two years of the misappropriation. If the misappropriator concealed the trade secret for the two year statute of limitations period, he was often able to use the trade secret as his own— thus permitting misappropriators a significant legal loophole.

At the federal level, the U.S. Congress also recently passed a law which provides stiffer criminal penalties against misappropriators of trade secrets.

Protecting Semiconductor Mask Works from Piracy: The Semiconductor Chip Protection Act

In the 1970s and early 1980s U.S. semiconductor companies had no intellectual property protection for their mask works. Mask works represent the complicated design pattern of a semiconductor chip. Mask work designs had been interpreted by the courts as normally not eligible for patent or copyright protection. U.S. semiconductor companies had spent millions of dollars to design mask works only to find that U.S. and foreign competitors were copying them at a cost of $50,000 or less to produce low-cost semiconductors. While the honest semiconductor companies had to recover their R&D on these mask works, the pirates could sell for less because they had much lower R&D costs.

To remedy the mask works piracy issue, the U.S. semiconductor industry participated in passing a new federal law. Called the Semiconductor Chip Protection Act of 1984, it created a new category of intellectual property protection for mask works. Companies could now protect their mask works after filing with the U.S. government for protection. The law gave them a ten year monopoly period for their mask works. Copiers could suffer heavy penalties and exclusion from the market. Reverse engineering is acceptable under the law, however.

Foreign mask work designers were not entitled to protection unless their home country offered similar reciprocal protection for U.S. companies doing business in their country. Today, most industrialized countries offer reciprocal protection, although some, like Korea, adopted it only recently. China has not yet adopted it. The World Trade Organization (WTO) Agreement on trade-related aspects of intellectual property (TRIPs) requires all WTO member countries to adopt such protection for mask works.

The public policy issue here is to continue to urge countries to adopt a chip protection act that is reciprocal to the U.S. act. Those countries which do not yet have a reciprocal act are beginning to realize that it is in their own self interest to participate.

SOUTH KOREAN WON

ANTITRUST AND INTELLECTUAL PROPERTY ISSUES

Collaborating with the U.S. Government to Protect Your Intellectual Property: "Special 301"

An important provision in the U.S. trade law to combat intellectual property piracy and infringement is "Special 301" of the U.S. Trade Act of 1974, as expanded by the Omnibus Trade and Competitiveness Act of 1988.

The purpose of "Special 301" is to identify foreign country policies and enforcement mechanisms that allow intellectual property piracy or infringement. The United States Trade Representative (USTR) is authorized to investigate on its own those countries lacking intellectual property protection and to initiate World Trade Organization (WTO) dispute settlement proceedings to address the problem. If the offending country is not a member of the WTO, then the USTR may file a Section 301 action to cure the problem. If the foreign country fails to heed a Section 301 finding of unfair trade practices, the USTR may impose trade sanctions against the foreign country.

"Special 301" requires the USTR to rank offending countries by category listings from serious violators to lesser offenders. The categories include: Priority Foreign Country, Priority Watch List, Watch List, and Other Observation Countries.

Lack of intellectual property protection in the countries on these lists has cost U.S. companies billions of dollars in lost sales and profits. USTR encourages such companies to help identify unfair intellectual property policies and practices of foreign countries and work with the USTR to find a solution.

As of the April 30, 1998 Annual Review, there was one Priority Foreign Country–Paraguay, 15 Priority Watch List countries, 32 countries on the Watch List, and 17 Other Observation Countries. China, a former Priority Foreign Country, is showing improvement and is in a special monitoring category.

Some of the most important "Special 301" Annual Review countries, as of April 30, 1998, along with their intellectual property violations, are as follows:

104 ___ BEYOND HIGH TECH SURVIVAL

Table 3.1 "Special 301" Annual Review April 30, 1998

PRIORITY FOREIGN COUNTRIES	ISSUE OF CONCERN
China	Removed to special monitoring category from Priority Foreign Country. Continuing problems of end-user piracy of business software, retail piracy, trademark counterfeiting, administrative protection for pharmaceuticals, trademark registrations and illegal importation of CD and VCD products.
PRIORITY WATCH LIST	
European Union (EU)	Denial of national treatment (same rights as EU companies), EU content requirements prejudicing U.S. companies, and trademark protection for pharmaceuticals.
India	Patent protection for pharmaceuticals and agricultural chemical products, general lack of patent and trademark law protection and lack of copyright enforcement. India was a Priority Foreign Country from 1991-1993.
Indonesia	Serious and continuing deficiencies in its law fostering software, book, video, VCD, drug and apparel trademark piracy; audiovisual market access barriers; and inadequate copyright, patent and trademark laws.
Italy	Ineffective anti-piracy legislation with regard to video, sound recordings and books, end-user software piracy, and restrictions on U.S. TV programming.
Russia	One of the largest pirate markets. Poor enforcement of intellectual property laws. No retroactive copyright protection for U.S. works and sound recordings.
Turkey	Copyright and patent laws inadequate. Use of illegal software in government offices.
WATCH LIST	
Canada	Copyright law discriminates against U.S. companies.
Hong Kong	Worsening piracy situation at retail level and for optical media.
Ireland	Inadequate copyright law contributing to high levels of piracy.
Japan	Lack of protection of software and trade secrets. End-user piracy of computer software. Inability to protect trade secret in court without disclosing it.
Korea	Limited retroactive protection for copyrighted works. Market access restrictions for motion pictures and cable TV programming. Lack of adequate protection of well-known trademarks, trade secrets and patents.

WATCH LIST *(cont)*

Singapore	Levels of intellectual property rights' protection the best in the Asia-Pacific region. CD-based copyright infringement on the rise.
Thailand	Piracy rates—particularly for videos and software—remain unacceptably high. Patent law inadequate. Illegal software use by government offices.
Brazil	Removed from Watch List for enacting modern laws to protect patents, computer software and copyrights. Piracy a problem. Difficult to obtain pharmaceutical patents.

OTHER OBSERVATION COUNTRIES

Germany	Audiovisual piracy. Pirating of "smart cards" and other "de-scrambling" devices used to steal encrypted satellite, cable and broadcast transmissions, particularly of U.S. motion pictures.
Mexico	Piracy and counterfeiting remain major problems. Poor enforcement.
Taiwan	Removed from Other Observation List. Pirating and counterfeiting of CDs, CD-ROMs and video games.

BRAZILIAN REAL

INDONESIAN RUPIAH

Using the Intellectual Property
Statutory Protection Periods

TYPE OF PROPERTY	STATUTORY PROTECTION PERIOD
Utility Patents	20 years from date of filing.
Design Patents	14 years from date of grant.
Biotech Patents	Special rule: Possible 1-5 year extension beyond the 20-year date of filing period when the patent application is subjected to long regulatory delays by the U.S. Patent Office.
Copyrights	Life of the author plus 50 years.
Trademarks	10 years from registration and indefinite life if continually renewed.
Trade Names	None. Look to contract and tort law.
Trade Secrets	None. Look to contract, tort and trade secrets law.
Know-how	None. Look to contract and tort law.
Chip Mask Work Designs	10 years from date of registration or first commercial exploitation, whichever comes first.

PORTUGESE ESCUDO

Summary

Laws passed by the U.S. Congress help companies save costs by permitting joint research and production activities among competitors. These laws have strengthened U.S. industry.

There is an emerging trend to identify and punish those engaged in unlawful monopolistic practices.

Section 337 and "Special 301" enable U.S. companies to protect their intellectual property from piracy and infringement.

Industry should continue to push for new laws to clarify the areas of submarine patents, copyright protection on the Internet, and the misappropriation of trade secrets.

AUSTRALIAN DOLLAR

INDIAN RUPEE

CHAPTER FOUR

FEDERAL TAXATION

KEY TOPICS

> *Maximizing Your Cash Flow through Accelerated Depreciation*

> *Aggressively Using the Research and Experimentation Tax Credit (Including State Analysis)*

> *Utilizing the Orphan Drug Tax Credit*

> *Understanding Employee vs. Independent Contractor Tax Issues*

> *Planning for Capital Gains*

> *Avoiding the Alternative Minimum Tax (AMT)*

> *Anticipating Alternative Tax Systems*

turning government policy into international profits ___ 109

FEDERAL TAXATION

This chapter covers some of the major federal public policy tax issues that impact company operations within the U.S. Depreciation is included because it is a cash flow generator. The research credit and the orphan drug credit, also cash flow generators, reward companies for their research spending. Employee versus independent contractor status is a critical issue for the software industry and for those companies that depend on temporary help. The capital gains issue is extremely important to investors in biotechnology, electronics and many other industries. I include a brief discussion of the Alternative Minimum Tax because it is a trap for the unwary. The controversial issue of simplified tax reform proposals for taxing individuals and companies is also discussed.

Maximizing Your Cash Flow
Through Accelerated Depreciation

The appropriate period (life) over which an asset can be depreciated has become an acute cash flow issue for some industries, particularly in the high technology sector. Many companies argue that technological obsolescence of their manufacturing equipment shortens the economic life of such assets and that U.S. tax depreciation rules do not keep pace with the rapid changes in the industry.

In fact, when compared to tax depreciation periods in other countries, the U.S. law is often not competitive. Companies in Japan, for example, report that they are able to depreciate some assets up to 88% in the first year while similar assets in the U.S. receive only about one-fourth as much in the first year.

When depreciation is too slow, companies pay more taxes

in the early years on income generated by such assets. Depreciation which is too fast will not match the tax deduction to the period when the income is generated. Congress has sometimes permitted faster (accelerated) depreciation to help companies reinvest in assets sooner as a way to spur the economy.

These accelerated depreciation tables are not perfect. Some industries have depreciation periods equal to one-half of their economic lives while other industries receive marginal benefit or have tax depreciation periods longer than the economic lives of their assets.

A case in point is the U.S. semiconductor industry. It has been very aggressive in its request for shorter depreciation periods. Arguing that its present five-year tax life has not kept pace with the current three-year economic life of such assets, the industry has introduced bills to shorten the tax depreciation period to three years. Congress has generally been sympathetic to a three-year life and has recognized the strategic defense and commercial importance of the industry. However, budget difficulties have prevented passage of a final tax bill that includes the three-year semiconductor depreciation period.

It has been over twelve years since Congress last overhauled the depreciation laws. In the view of many industries, the time is right for another overhaul.

Aggressively Using the Research and Experimentation Tax Credit (Including State Analysis)

The Research and Experimentation (R&E) tax credit was enacted in 1981 to address the perceived problem of U.S. industry's under-investment in research. A tax credit (initially at a 25% rate) was applied to increased R&E spending over a prior three-year R&E spending average. In 1989, the credit formula was modified to more closely follow a company's sales growth.

One of the main issues with the credit is that it is not permanent. Historically, the R&E tax credit has been passed for only one to two years at a time, thus denying R&E intensive companies any certainty as to its benefit. The credit formula is quite complex, and proving the amount of the credit to the satisfaction of IRS auditors has been a challenge for some companies. But the credit is important in lowering a company's tax rate, and hence its bottom line net income and earnings per share.

The credit, also commonly known as the Research and Development (R&D) Tax Credit, applies to U.S.-based research (essentially wages and supplies attributable to R&E activities) performed in a company or through payments to universities to perform the research. The current credit rate is 20%. To avoid a doubling up of a tax credit and a business deduction for the same R&E spending, the amount of the credit is subtracted from the total R&E spending amount to provide a net R&E expense deduction. Many states, including California, have enacted R&E credits along the federal model. Foreign corporations doing R&E in the U.S. may also be eligible for the credit.

The 1989 R&E credit formula rewards R&E spending that exceeds a prior four-year average base amount. The base amount is determined by multiplying an R&E percentage (the percent of R&E to sales for the five years 1984-88) by a company's prior four-year average gross receipts. The maximum base amount allowed is 16% of gross receipts. Then, to the extent current year R&E exceeds the base amount, a 20% credit rate is computed.

COLUMBIAN PESO

Applying this formula to hypothetical numbers, assume the following:

R&E to sales percentage for 1984-88	10%
Prior four-year average gross receipts	$1,000
Current year R&E spending	$150

Thus:

Current Year R&E	$150
Less base (10% x $1000)	–$100
Incremental R&E	$50
Tax credit (at 20% rate)	$10

In the case of start-up companies, the base is 3% for the first five years and is phased in to the normal base formula over the second five-year period.

Some states, like California, which follow the federal law, will determine the base using only local (state) R&E and gross receipts, with a 16% maximum R&E to gross receipts base amount. Thus, in the example above, if the California R&E to gross receipts percentage were 50% for the years 1984-1988, California average gross receipts for the prior four years were $200, and current year California R&E were $150, then the credit would be computed as follows:

Current Year R&E	$150
Less base (16% x 200)	–$32
Incremental R&E	$118
Tax credit (at 11% rate)	$13

A major issue for some high R&E spending biotech companies is a provision in the law to limit the R&E eligible for the credit to no more than twice the base amount. Thus, if a company has current year R&E of $300 and a base of $100, the base would need to be increased to $150 to satisfy the test, thus limiting eligible R&E by $50 and reducing the credit by $10 ($50 x 20%).

Some R&E-intensive electronics companies have complained that the current federal credit formula does not give them enough incentive to do R&E. In fact, some of the highest R&E spenders in terms of sales percentage and absolute dollars have not received a credit for several years. To provide an incentive to these companies, Congress recently passed an optional formula to reward high R&E spenders who do not benefit under the current formula. The optional formula may be used if a company so elects on an irrevocable basis. The normal 20% credit drops to a maximum credit of 2.75% on that portion of current year R&E spending in excess of 2% of the prior four-year average gross receipts. Thus, if the prior four year average gross receipts were $1,000, and current year R&E spending $150, then the R&E base amount would be $20 ($1,000 x 2%), and R&E in excess of $20 eligible for a credit would be $130, and the credit would be $3.58 ($130 x 2.75%).

Tax incentives such as the R&E tax credit highlight the savings opportunities for companies. There are new proposals to extend this credit and to increase the maximum credit percentage for the optional formula. Astute management will recognize these opportunities and hire competent and imaginative tax professionals to capture these incentives. These companies will lower their tax burden, increase their cash flow and enhance their earnings per share.

Utilizing the Orphan Drug Tax Credit

The orphan drug tax credit is important to the biotech industry. A 50% tax credit applies to clinical testing expenses for developing unprofitable drugs for rare diseases. Generally, the testing must take place in the U.S. Unlike the R&E credit, it is permanent and there is no prior year base that must be exceeded in order to qualify for the credit.

The term "clinical testing expenses" refers to wages and

supplies attributable to the testing. Some equipment rentals and contract research expenses paid to non-employees also qualify as clinical testing expenses. A rare disease generally is one afflicting fewer than 200,000 people in the U.S. Diseases considered rare in the U.S. include Lou Gehrig's disease and muscular dystrophy.

Without the financial assistance the credit provides, research in this area may suffer, as it is uneconomical for many biotech companies to embark on such endeavors.

Understanding Employee vs. Independent Contractor Tax Issues

One of the major tax issues for companies is the definition of an individual as an independent contractor or an employee. The issue is particularly acute for those companies hiring software designers, computer programmers, and temporary or part-time helpers. The issue is important from both a federal and state tax perspective.

Microsoft Corporation recently felt the pain of this difficult issue when a federal court ordered it to provide retroactive profit-sharing and stock options for certain of its "freelance" or contract workers who had been hired as independent contractors.

Many employers claim the current tests as set forth by the Internal Revenue Service (IRS) are not working. In these tests, the IRS attempts to determine whether an individual comes closer to independent contractor status or employee status. The common law test for an independent contractor is one that looks at whether the individual has control over the means of getting to the final result vs. whether someone else controls both the means and the result. If an individual has control over the means, he is an independent contractor. If he does not have such control, he is an employee.

The IRS uses a multiple test to make this determination. Important factors include setting hours of work, doing work off-site, payment by the hour, week or month, payment of business and traveling expenses, furnishing tools or materials, realization of profit or loss, working for more than one firm, and making services available to the general public.

Employers run the risk that if they improperly classify an individual's status as an independent contractor, the IRS will impose large fines as well as collect employment taxes. The employer will also be liable for the employee's social security taxes, federal and state income tax withholding, and unemployment taxes.

Employers have been leading an effort to clarify the definition of independent contractor status. Safe harbor provisions should be established which would protect the hiring firm from penalties, and firms should be relieved from back taxes for mis-classification when form 1099s are filed and there is no evidence of fraud.

Because of aggressive and coordinated efforts by employers, Congress recently has become more active in helping provide the classification necessary to differentiate an independent contractor from an employee. The business community, however, needs to continue to put money and time into this lobbying effort to achieve even further gains.

Planning for
Capital Gains

Reduced federal and state tax rates on capital transactions have been on again and off again for many years. Capital gains tax reduction is a major issue for some industries and their employees.

Currently, the federal capital gains tax rate for individuals generally is 20%, as opposed to the highest rate on ordinary income of 39.6%. States vary as to whether individuals should get a capital gains tax reduction. Corporate federal capital gains are taxed at 35% (the same rate as corporate ordinary income). Federal and state corporate capital gains tax reductions are issues of little importance for many companies. To individual shareholders of these companies, however, a capital gains tax reduction is usually very important and can make the difference whether the individual makes or doesn't make the investment.

The rationale for a lower capital gains tax rate is that investors would be more willing to invest in startups and other risky companies if they could enhance their return on investment with a lower capital gains rate.

Opposition to the capital gains tax reduction has come from consumer groups and unions who claim the capital gains reduction is a boon to the rich. Federal lawmakers have been divided on its merits. The division is often along party lines, with Republicans favoring and Democrats opposing the capital gains reduction.

Economists are also divided. Some adhere to a conservative static model approach that looks purely at the revenue loss from the lower rates—increased economic activity from the lower rates is not taken into account. Other economists look at capital gains reductions using a dynamic model approach which might conclude that a lower capital gains rate would stimulate economic activity and more sales of capital assets, such as real estate and stocks, and raise revenue because of the increased activity.

The debate over whether capital gains reductions will lower or increase tax revenues will go on for many years. Today supporters of lower capital gains rates are the winners. Will lower capital gains rates stay in the law? We shall see.

Avoiding the Alternative Minimum Tax (AMT)

One of the income tax provisions that surprises many companies is the Alternative Minimum Tax (AMT). It can create a tax liability for a company when most business people would think an income tax would not apply. The AMT can also have unexpected tax consequences for a company's employees who exercise Incentive Stock Options.

The AMT was passed to curb a perceived problem that some companies were reporting losses and not paying any income tax, yet reporting earnings on financial statements. The same was true for certain individuals. As a result of tax shelter programs, tax-exempt interest vehicles and a lower capital gains rate, many of these companies and individuals were in fact not paying taxes.

The concept of the AMT is simple. Companies and individuals should not be able to use tax incentives to eliminate their tax liabilities. Companies and individuals should be expected to pay some tax if they earn true economic income during the year. The AMT tax rate is 20% for companies and up to 28% for individuals on alternative minimum taxable income. Alternative minimum taxable income is taxable income plus tax preference items (what I call AMT items).

While the concept of AMT is simple, the application is not. Some AMT rules go beyond taxing true economic income. For example, if the economic depreciation period for your company's equipment is shorter than its tax depreciation period, your company may pay more AMT taxes than companies with economic depreciation periods approximating their tax depreciation periods.

The important basics of the AMT for companies and individuals include the following areas:

> Tax exempt interest—The amount of tax exempt interest on certain bonds is an AMT item.

> Net operating losses—Only 90% of such losses are allowable, meaning the remaining 10% cannot be used to reduce AMT items currently, but may be used as a carry forward.

> Foreign tax credits—Only 90% of such credits are allowable, meaning the remaining 10% cannot be used to reduce AMT tax currently, but may be used as a carry forward.

> Incentive Stock Options—The difference between the fair market value of shares at exercise of the option over the option price (gain element) is an AMT item.

The AMT income adjustments can be reduced by corporate and individual exemption amounts. The corporate amount is $40,000 and the individual amount ranges from $22,500 to $45,000, depending upon the individual's filing status. These exemptions phase out to zero after AMT income reaches certain levels.

The above represents some of the basic AMT items, deductions and exemptions. There are more. If you think you are getting a tax benefit for certain tax vehicles, look again as the AMT rules may cut your benefit in half.

This is an area where companies and individuals may need professional help. The accounting firms may be the most helpful here with their tax preparation expertise and sophisticated computer planning programs.

There have been industry efforts to pare down or eliminate the AMT on the grounds that it is administratively too burdensome and unfair. My personal view is that it is both of these. Industry lobbying of the Congress and the Administration could prove useful to reduce the administrative burdens and complexity of the AMT. Moreover, several of the tax shelter programs and tax schemes allowing significant tax reductions have been eliminated by other changes in the tax law. Some say the goals of the AMT have therefore been satisfied and that there no longer is a need for the AMT in the law.

Anticipating Alternative Tax Systems

There is a movement in the U.S. to adopt a fairer, simpler tax code to encourage investment and savings and to discourage consumption. Proponents of such a tax system believe the current Internal Revenue Code is unfair, too complex, and has too many provisions encouraging consumption.

The main alternative tax systems under discussion are the flat tax and the value added tax. What the two systems have in common is that they tax consumption and encourage savings and investment. They differ in their complexity. However, they are simpler than the existing system and generally have tax rates under 20% for businesses and individuals (with a few exceptions).

They allow 100% expensing of capital assets and abolish the alternative minimum tax. While they do not tax investment income and capital gains, they disallow certain deductions (interest expense, state and local income tax deductions) and credits (R&E, orphan drug, and foreign tax credits).

Most of the discussions of these alternative tax proposals have been directed at the individual level. There are many outstanding conceptual issues at the corporate level that need to be addressed. There will be winners and losers, depending on such factors as whether the companies are highly debt-leveraged, are labor intensive, have high capital expenditure growth, are importers or exporters, or have high or low profit margins. Long transition rules from one system to the next are essential to help soften the impact on different industries.

AUSTRIAN SCHILLING

IRISH PUNT

Summary

Even if your company is fortunate enough to make a pre-tax profit, U.S. tax laws could leave you with little to show for your efforts. Use of accelerated depreciation for your plant and equipment and research tax credits will increase your after-tax cash flow.

The issue of whether a worker is an independent contractor or an employer has been an annoying, if not a costly problem for many companies. Legislation is needed to address this problem.

The reduced capital gains tax rate is an attractive incentive for investors. The business community should work to keep this incentive in the law for a long time.

The Alternative Minimum Tax should be repealed. Industry needs to keep up its lobbying efforts in this area.

Simplifying the U.S. tax laws is a worthwhile goal. Flat-tax and value-added tax proposals have appeal to many in industry. Critical issues will include the form of the tax change and transitional relief periods. This area requires thorough debate and it is unlikely that we will see a new tax system in the 1990s.

THAI BAHT

FINNISH MARKKA

SINGAPORE DOLLAR

INTERNATIONAL TAXATION

KEY TOPICS

> *Taking Full Advantage of the Foreign Tax Credit*

> *Lowering Your Tax Rate on Exports: Using the Foreign Sales Corporation*

> *Aggressively Deferring Taxes on Offshore Operations*

> *Selling to Your Overseas Subsidiaries: Being Aware of Transfer Pricing Issues*

> *Manufacturing in U.S. Possessions: The U.S. Possessions Income Tax Credit (Section 936)*

> *Taking Advantage of Tax Treaties*

INTERNATIONAL TAXATION

Taxation of overseas operations is an arcane but important subject for companies. Government policy can impact these operations by causing double taxation. With a thorough understanding of international tax issues, double taxation can be avoided, taxes lowered, and overseas manufacturing operations structured to allow overseas companies to build up cash free of current U.S. and foreign taxation. The Foreign Sales Corporation, the Possessions Tax Credit, and certain tax treaties can also reduce the tax bill of U.S. companies with offshore operations.

Taking Full Advantage
of the Foreign Tax Credit: *General Principles*

The U.S. foreign tax credit (FTC) is important to a company with income subject to taxation both in a foreign country and in the U.S. The purpose of the FTC is to prevent double taxation of the same income by allowing a foreign tax liability to be offset dollar for dollar against the U.S. tax liability on the same income. A company's ability to claim an FTC will have a significant impact on its bottom line net income.

For example, if a U.S. company doing business in Taiwan through a branch office incurs a 25% tax rate on $1,000 of income at the branch, the resulting $250 of tax ($1,000 x .25) may be offset against the $350 of U.S. tax incurred on the same branch income ($1,000 x .35), and the net U.S. tax will be $100 (U.S. $350 less Taiwan $250).

U.S. taxation applies to a U.S. company's worldwide income. Thus, the FTC will come into play not only in the foreign branch income situation above, but also when income is received from abroad in the form of dividends, interest, royalties, rents, and other gains.

The FTC is exceedingly complex. Entire books have been written about it. In general, the important issues of the FTC consist of the following:

> Whether the foreign tax is based on income (only taxes on foreign income can be creditable).

> The amount of foreign source taxable income.

> The amount of worldwide taxable income.

> The amount of U.S. tax owed before a foreign tax credit has been claimed.

Looking more closely at the above four issues, the following observations are appropriate:

> **Foreign income tax.** Generally this would not include a sales tax, a property tax or a gross receipts tax.

> **Foreign source taxable income.** This is the U.S. company's taxable income after subtracting foreign source expense from foreign source gross income.

> **Worldwide taxable income.** This is the U.S. company's worldwide gross income less worldwide expense.

> **U.S. tax before FTC.** This is the amount of U.S. tax liability before applying the FTC.

One important consideration in claiming the FTC is that the FTC cannot exceed what the U.S. tax liability would have been on the same foreign income. This limitation is accomplished by the following calculation:

UNITED ARAB EMIRATES DIRHAM

$$\frac{FSTI}{WTI} \times \text{U.S. tax liability} = \text{maximum amount of creditable foreign taxes}$$

Where: FSTI = foreign source taxable income
WTI = worldwide taxable income
U.S. tax liability = the U.S. tax before FTC

Applying this limitation principle to the Taiwan branch situation above, the full calculation of the foreign tax credit would be as follows:

U.S. tax return income from Taiwan branch	$1,000
U.S. taxable income from U.S. operations	$1,000
Worldwide taxable income	$2,000
U.S. tax (35% rate) on $2,000	$700

1. Maximum amount of FTC that can be claimed:

$$\frac{FSTI\ \$1,000}{WTI\ \$2,000} \times \$700 = \qquad \$350$$

2. Amount of foreign tax paid $250

3. Remaining U.S. tax liability   ($350–$250) $100

If foreign taxes paid exceed the maximum amount of the FTC that can be claimed because of the limitation principle above, the law provides a two-year carryback and a five-year carryforward period to use up the excess foreign taxes. Foreign taxes not used in this window will be lost forever and double taxation will result to the extent of the unused foreign taxes.

Over the years, there have been proposals in the Congress to eliminate or cut back on the benefits of the FTC. A loss of all or part of the FTC would be a serious bottom-line blow to industry. It must fight aggressively such attempts in the Congress.

The amount of the usable foreign tax credit can be increased in several ways as shown in the next topic.

Enhancing Your Foreign Tax Credit:
The Treatment of Foreign Source Income and Expenses

Note that the greater the ratio of FSTI to WTI, the greater will be the amount of eligible FTC. From this observation, the following maxims hold:

> Generate as much foreign source income as possible.

> Keep to a minimum foreign source expense.

> Keep to a minimum U.S. source income.

> Generate as much U.S. source expense as possible.

Remember, the goal here is to use all of your foreign tax credits. With careful planning you can increase your foreign source taxable income and accomplish that objective. Some techniques used by companies to generate foreign source income include the following:

> Earn dividends, interest, and royalties from foreign, rather than U.S., sources.

> Sell from the U.S. to overseas locations with title passage outside the U.S. (such as F.O.B. Munich).

> Make certain your selling price to an overseas subsidiary results in the appropriate amount of profit to the subsidiary.

Ways to lessen foreign source expense include:

> Keep foreign source expenses, such as interest and royalty expenses, to a minimum.

> Properly utilize the R&E allocation to foreign subsidiaries (one of the most hotly contested public policy tax issues today). The IRS requires a certain amount of U.S. R&E expense to be apportioned to foreign source income based either on the ratio of foreign to worldwide sales or foreign to worldwide income. As an export tax incentive, some U.S. R&E is allocated first to U.S. source expense. The IRS believes a 30% allocation is appropriate. Over the years, Congress has advocated increasing the allocation to U.S. source expense in the 50-67% range. Currently it is 50%. Again, obviously, the greater the expense allocation to the U.S., the greater will be foreign source income.

It also goes without saying that you will best utilize your

foreign tax credits if your foreign tax bill is kept to a minimum. This can be done by locating your foreign operations in low- and zero-tax countries, using tax treaties wisely, and keeping an eye on the appropriate transfer price between the U.S. parent company and its foreign subsidiaries. As noted in the section on tax treaties, your company may be able to use a tax treaty to completely avoid paying income taxes to the foreign country if you are able to structure your overseas operations not to be doing business in that country.

Lowering Your Tax Rate on Exports:
Using the Foreign Sales Corporation (FSC)

The Foreign Sales Corporation (FSC - pronounced "fisk") is an important U.S. public policy export incentive. The FSC and its predecessor, the Domestic International Sales Corporation (DISC), have been in the law since 1971 for the purpose of stimulating exports from the U.S.

DISC and FSC were put in the law to counter the tax advantage European companies gained through a forgiveness of value added taxes for their exports from the European Union.

FSC reduces a U.S. company's federal income tax rate on qualifying export transactions from 35% to under 30% (and sometimes down to the middle teens and low twenties). This rate reduction is characterized as a permanent accounting difference under the Financial Accounting Standards Board (FASB) rules, and thus can increase bottom line and earnings per share considerably—a benefit for the company's financial statement.

Usually, a FSC will do business as a commission agent or on a buy-sell basis. Its commission or sales profit will be the amount determined under the FSC incentive rules. To satisfy WTO (formerly GATT) rules, the FSC must be incorporated outside the U.S. (Guam and the U.S. Virgin Islands are among the more common qualifying places of incorporation), and have some overseas corporate substance. Most FSCs, for simplicity of operation, do business as a commission agent.

The FSC rules determining what qualifies as a U.S. export n the calculation of the FSC benefit are complex. Generally, a qualifying U.S. export is one derived from the sale or lease of property manufactured, produced, grown, or extracted in the U.S. An export also includes services related to such property.

The FSC tax benefit is calculated according to a variety of methods, but the two most common methods are shown in Table 5.1. This simple example shows how the FSC incentive works with the 15% of profit exemption and the 1.19% of sales exemption. The taxpayer may use whichever method produces the greater result on each individual export sale. Loss transactions may be excluded in determining the FSC incentive, and this will result in lowering the FSC's tax rate on exports even more.

Table 5.1 Most Common Methods of Calculating FSC Tax Benefit

CATEGORY	15% OF PROFITS EXEMPTION	1.19% OF SALES EXEMPTION
Sales from exports	$ 2,000	$ 2,000
Pre-tax export profits	200	200
Exemption from taxable income	30	23.80
Taxable income after FSC benefit	170	176.20
Tax (at 35% rate)	$ 59.50	$ 61.67
Effective FSC tax rate on qualifying export profits of $200 (tax paid divided by$200)	29.75%	30.84%

Smart companies put considerable effort into maximizing the FSC benefit by consulting with competent tax lawyers and accountants and by using sophisticated computer programs to capitalize on every possible FSC nuance.

Since its inception in 1971, Congress and the Internal Revenue Service have attempted to reduce the benefits of the FSC through law changes or onerous audits. Occasionally, there have been efforts to repeal the FSC provisions.

In late 1997, an assault on the FSC was brought by the European Union in the WTO. The case was precipitated by competitive tensions between Airbus and Boeing. European high tech companies were also alarmed that the U.S. Congress passed an amendment to the FSC law in 1997 to give FSC benefits to software companies like Microsoft and Oracle. The European Union filed the case on the basis the FSC was an illegal export subsidy forbidden by WTO rules. As of this writing, no decision has yet been rendered.

As exporting is important to the bottom line of most companies, industry must work hard to preserve the FSC benefit.

Aggressively Deferring Taxes on Offshore Operations: *The Concepts of Controlled Foreign Corporations and Subpart F Income*

Prior to 1962, income of a foreign subsidiary of a U.S. parent company generally was not taxed by the U.S. until such time as the foreign subsidiary paid a dividend to the U.S. parent. This U.S. tax law was amended in 1962 to curb a perceived abuse that U.S. companies were trying to avoid U.S. tax by setting up foreign subsidiary tax haven companies and transferring income that would otherwise be taxed in the U.S. to the foreign subsidiary companies.

The 1962 amendments set up a whole new body of U.S. tax law which involves the concepts of a controlled foreign corporation and Subpart F income.

A controlled foreign corporation is typically one where the U.S. parent company owns more than half of its stock. Subpart F income stems from a section in the Internal Revenue Code under the heading of Subpart F which lists several categories of income of a foreign subsidiary ("tax haven" income) and subjects such income to current taxation in the U.S., even if the subsidiary has not paid a dividend.

To have a situation where Subpart F income exists, there must first be a controlled foreign corporation relationship.

Once such relationship exists, certain classes of income of the controlled foreign corporation will be subject to current U.S. taxation. This includes many types of passive income, such as dividends and interest, and certain types of operating income arising from sales or services.

A typical type of Subpart F sales income occurs when a U.S. parent sells a product to its Hong Kong distribution subsidiary which then resells the product to customers in various Asian countries. The income of the Hong Kong subsidiary, assuming it is more than 50% owned by the U.S. parent company, would be subject to current taxation in the U.S. in an amount equal to the U.S. shareholding percent times the income.

Perhaps the most important exception to the Subpart F rules occurs in the case of the foreign selling subsidiary which can prove that it is a manufacturer. If sales income for the subsidiary comes from manufacturing, its income will not be Subpart F sales income, and current U.S. tax will not apply. Manufacturing exists where the foreign subsidiary engages in substantially transforming the product (e.g., fabricating semiconductors), is engaged in a number of steps generally considered to be manufacturing (e.g., building a car), or has local country costs equal to 20% or more of the product's cost of production in its foreign subsidiary's activities.

Periodically, in budget balancing efforts, Congress has attempted to reduce or eliminate ways to avoid the Subpart F area. One such attempt has been to tax the income of so-called "runaway" plants (i.e., business operations moved from the U.S. to overseas companies). Industry arguments against these actions include the following:

> Hardly any other country has the concept of Subpart F income.

> U.S. companies would be put at a competitive disadvantage with foreign competitors from the standpoint of cash flow and bottom line net income.

> The tax revenue raised is not significant.

With careful planning based on a thorough understanding of Subpart F principles, a U.S. parent company will benefit from increased cash flow, lower taxes, and greater net income.

Selling to Your Overseas Subsidiaries:
Being Aware of Transfer Pricing Issues

International transfer pricing among related companies is more of an art than a science. Transfer pricing represents a delicate interplay between the U.S. parent and its foreign subsidiaries that involve the following considerations:

> The correct amount of U.S. profit allocation.

> The correct amount of foreign subsidiary profit allocation.

> The correct duty payments in the U.S. and foreign countries.

> The best utilization of foreign tax credits.

> The avoidance of U.S. and foreign country fines and other penalties.

The standard method of pricing for tax and duty purposes is called arms length transfer pricing (i.e., transfer pricing between related companies that would be comparable to pricing between unrelated companies). Under U.S. law, Section 482 of the Internal Revenue Code, arms length transfer pricing between related companies may involve several different acceptable methodologies. But the U.S. consequence is only one small part of a complicated web of intertwining transactions.

For example, if the transfer pricing results in a profit allocation acceptable to U.S. tax authorities, it may not be acceptable to foreign tax authorities or to U.S. or foreign customs. If a transfer price profit allocation is acceptable to a foreign tax authority, it may not be acceptable to the U.S. IRS or to a foreign customs authority.

Further, the amount of profit allocated to a foreign subsidiary may have a direct bearing on whether foreign tax credits can be utilized on the U.S. tax return. Finally, if the profit allocations are viewed as too insufficient or as fraudulent, fines and perhaps even criminal penalties could apply.

Here is a hypothetical case:

A U.S. manufacturer sells to its foreign subsidiaries in England, France, Germany, and Japan. Its U.S. cost of production is $1.00 per unit. It sells to its subsidiaries in England,

France, Germany and Japan for $2.00 per unit, resulting in U.S. profit per unit of $1.00. Its foreign subsidiaries sell to end customers in their respective countries at the following prices: England–$3.00; France–$1.00; Germany–$5.00; and Japan–$10.00. These transactions are as follows:

	U.S.	ENGLAND	FRANCE	GERMANY	JAPAN
Sale Price to Customer	$2	$3	$1	$5	$10
Cost of Sales	$1	$2	$2	$2	$ 2
Pre-tax Profit (loss)	$1	$1	$(1)	$3	$ 8
Duty Value	NA	$2	$2	$2	$ 2

Appearance of reasonableness of duty value and pre-tax profit to government auditors:

	U.S.	ENGLAND	FRANCE	GERMANY	JAPAN
U.S. IRS	OK	OK	Great	More profit should be allocated to U.S.	U.S. transfer price profit fraudulently low
Foreign Customs	NA	OK	Great	Duty value too low	Duty value fraudulently low
Foreign Tax Authorities	NA	OK	Transfer price too high	Great	Wonderful allocation

GERMAN MARK

As you can see, appearances as to the appropriateness of transfer prices and profit allocations can vary from country to country. The key in each of these countries was the end-customer selling price. How could these end-customer selling prices vary so much from country to country? Possible explanations might include:

ENGLAND	Transfer pricing and profit allocation show normal market conditions.
FRANCE	The French subsidiary was trying to penetrate a new customer base with attractive prices, or it had excess inventory and was trying to unload it.
GERMANY	Market demand conditions are strong and pricing reflects this high demand, or the sale was of a small quantity bringing a premium price (a sale of a large number of units might have been made at $3.00-$4.00 per unit).
JAPAN	Market demand looks extremely high, or the subsidiary has no competition in this market, or the price was set at a premium for an extremely small sales amount.

As transfer pricing involves many U.S. and foreign tax and customs variables, it is critical that you have the help of U.S. and foreign tax and customs consultants and a capable in-house staff able to monitor, usually through sophisticated computer programs, the optimal tax, customs and cash flow transfer pricing techniques.

A major public policy issue here is that our own U.S. government and foreign governments (through tax treaties) arguably should provide sufficient transfer pricing guidelines for companies. Recently, the U.S. IRS has been doing just this, using a procedure called Advanced Pricing Arrangements (APA). Such arrangements are a step in the right direction to help honest companies avoid getting into legal trouble.

Manufacturing in U.S. Possessions:
The U.S. Possessions Income Tax Credit (Section 936)

The Possessions Tax Credit is an incentive to help lower the unemployment rate in U.S. possessions (Puerto Rico, Guam, U.S. Virgin Islands, and Western Samoa). If a company qualifies for the tax credit, no further U.S. tax applies. Companies in the pharmaceutical and electronics industries have been active in taking advantage of this benefit.

Qualification for the credit (the so-called Section 936 credit) requires strict adherence to a complex recipe of tax rules. These rules require that a qualifying company be incorporated in the U.S. and earn a certain amount of income from active manufacturing operations within the possession. Passive income, such as interest, must generally be earned within the possession as well.

The Section 936 credit works well when the possession itself offers a tax exemption period or a reduced tax rate for a period of years. Failing these tax incentives, operating in a possession may be less attractive. Other benefits of operating in a possession, however, might include a lower wage structure (although a company must pay the U.S. minimum wage to avail itself of Section 936), and zero duties on shipments to the U.S. (possessions are considered within the U.S. customs territory for this purpose).

To establish a Section 936 company, you should consult a competent U.S. international tax lawyer or accountant and experts practicing law and accounting in the U.S. possession.

Congress, in recent years, has been reducing the benefits of Section 936 companies. Opponents argue that Section 936 has served its purpose. It has brought down unemployment in the possessions and should now be eliminated entirely to reduce the budget deficit or to provide funding for other tax provisions.

Taking Advantage of Tax Treaties

The U.S. has entered into income tax treaties with many of the European and Asian countries as well as with Canada and some Caribbean countries. Tax treaties generally make it easier for U.S. companies to lower their income tax liabilities in the foreign treaty country.

The test of what activities will be considered to be "doing business" in a foreign treaty country will often be more beneficial to U.S. companies than the internal law of the foreign country. In other words, a tax treaty allows a company to sell more in the foreign country before being found to be engaged in a taxable business than it could if there were no tax treaty.

Taxes imposed by the foreign treaty country on certain income, such as dividends, interest, and royalties paid to U.S. companies, are generally at lower rates than payments to companies in non-treaty countries. In addition, tax disputes between the U.S. company and the foreign treaty country are easier to administer because of treaty provisions designed to ensure that the U.S. and the treaty country do not doubly tax the same income.

For many companies, a major U.S. public policy issue involves the absence of so-called tax sparing provisions in U.S. treaties. These tax sparing provisions are commonly found in European and Japanese tax treaties with developing countries, such as Malaysia, Indonesia, and Thailand.

The U.S. does not have tax sparing provisions with developing countries. The net result is that the U.S. company ends up at a competitive disadvantage to the European and Japanese companies. Tax sparing provisions make it possible for European and Japanese companies to invest in Southeast Asia on essentially a tax-free basis both there and in their home country, while a U.S. company must pay tax in the U.S. on income from such countries.

This competitive disadvantage works as follows: Assume a U.S. company (USA) and a Japanese company (Japan) each invest in a semiconductor assembly plant in Malaysia. Malaysia offers USA and Japan 10-year tax exemptions from payment of Malaysian income taxes on income from the assembly plants. Without the tax exemption, the normal Malaysian tax would be at a 40% rate. When USA pays a dividend from Malaysia, the normal 35% U.S. tax will apply. When Japan pays a dividend from Malaysia, virtually no Japanese tax will apply because the Japanese tax treaty permits a hypothetical Malaysian tax credit at a 40% rate on the dividend distribution. Japanese taxes apply only if they exceed the 40% rate. Clearly, the U.S. Congress should be urged to change this unfair competitive disadvantage for U.S. companies.

The area of tax treaties is one of specialization. A competent international tax lawyer or accountant is essential to maximize the optimal planning benefits of U.S. and foreign income tax treaties.

MALAYSIAN RINGGIT

CUBAN PESO

Summary

The most important international tax issue for U.S. multi-nationals is double taxation (taxation in the foreign country as well as in the U.S.). The foreign tax credit helps reduce double taxation.

Cash flow resulting from overseas manufacturing operations can be protected from current U.S. taxation through prudent use of the U.S. tax laws. Cash flow can also be generated through legal transfer pricing techniques.

For both large and small exporters, the Foreign Sales Corporation may allow you to dramatically reduce your U.S. tax burden on exports.

Wise use of the foreign tax credit, tax deferral rules, and the Foreign Sales Corporation will also have a significant impact on your financial statement bottom line.

SOUTH AFRICAN RAND

EMPLOYEE COMPENSATION ISSUES

KEY TOPICS

> *Helping Provide for Your Employees' Retirement:*
> *Profit-Sharing, 401(k) and Pension Plans*

> *Understanding Stock Purchase Plans*

> *Incentivizing Your Key Employees: Stock Option Plans*

> *Using Cafeteria Plans*

> *Using Employee Tuition Reimbursement Plans: Section 127*

EMPLOYEE COMPENSATION ISSUES

This chapter discusses the most common benefit and compensation plans offered to employees. These include retirement plans, stock purchase and stock option plans and some common employee reimbursement plans.

Helping Provide for Your Employees' Retirement: *Profit-Sharing, 401(k) and Pension Plans*

Retirement plans are a very important part of the culture of companies. As employees near middle-age, they begin to express more interest in planning for retirement.

The most common type of retirement plan for companies is the profit-sharing plan. This plan is found in Section 401 of the Internal Revenue Code and is legally described as a defined contribution plan. A profit-sharing plan fits well in a culture of risk-reward mentality. If a company makes a profit, it will contribute some of its profits to its employees in the profit-sharing plan. Logically, if the company makes no profits, it follows that no contribution to the profit-sharing plan will be made. Start-up companies, in particular, are hesitant to use precious cash resources for retirement plan contributions when they are not making profits, and thus prefer profit-sharing plans.

Within the past 10-15 years, a newer plan has sprung up which also derives from Section 401 of the Internal Revenue Code. Called the 401(k) plan, it encourages employee retirement savings and provides a complement to profit-sharing plans. The purest 401(k) plan involves no company contributions. Employees contribute a certain percentage of their pre-tax profits (up to a maximum approaching $10,000) to the 401(k). Since the employee's contribution comes from pre-tax wages, his contribution reduces his taxable wages and thus his tax bill. Of companies offering 401 (k) plans, over 90% match employee 401(k) contributions by contributing to the 401(k) a percentage of the employee's contribution or an amount based on after-tax profits or some other corporate performance measure. Tying contributions to corporate performance is a rapidly growing trend. About 90% of the corporate community has 401(k) plans.

The third type of plan is the pension plan, which also stems from Section 401 of the Internal Revenue Code and which is legally described as a defined benefit plan. In this plan, the company typically will contribute a percentage of the employee's salary to the plan, whether the company has been profitable or not.

As retirement plans are important to their employees' well-being and to company cost containment, companies will want to watch public policy proposals impacting them.

Understanding
Stock Purchase Plans

Section 423 of the Internal Revenue Code provides a way for companies to raise capital by establishing plans permitting employees to purchase the company's stock at a discount. Many companies have such a plan.

Typically, the employee, through payroll withholding, can purchase company stock at a 15% discount off the lower of the beginning or end of the period's stock price. If a two-year holding period is met after purchase of the stock, the employee can receive a lower capital gains tax rate on any gain from the sale. If the holding period is not met, the employee will be taxed at ordinary income tax rates.

A typical plan works as follows: Assume employee contributions through payroll deductions of $1,000 over a plan period of 3 months. Assume further that the stock price at the beginning of the period was $20 and at the end of the period was

$30. Applying the 15% discount on the lower of the $20 or $30 price gives the employee a purchase price of $17 per share, for a total of 59 shares on $1,000 of payroll deductions. If the employee sells his shares immediately after the end of the period, he will have a gain of $13 per share ($30 less $17) taxed at ordinary rates. If he holds the shares for two years at a time when the shares are selling for $50 each, he will have a capital gain of $33 ($50 less $17).

Congress will probably keep these plans intact as a way to help companies raise needed capital and for employees to have an ownership stake in their company as they plan for their retirement.

Incentivizing Your Key Employees: *Stock Option Plans*

Stock options represent an employee benefit that aligns the employee's interest in the success of his company with the company's shareholders. If the company performs well, the employee should also benefit through stock appreciation.

The most common stock options are Nonqualified Stock Options (NSOs) and Incentive Stock Options (ISOs). ISOs are arguably better than NSOs when a low capital gains provision is in the law. The counter-argument in favor of NSOs is that the company is entitled to a wage deduction on its tax return equal to some of employee's stock gain, a feature not permitted in an ISO. The company gets the deduction even though it never actually paid the wages to the employee.

A burning issue for employees with ISOs is the application of the Alternative Minimum Tax (AMT) at the time the employee purchases the stock. Originally a tax provision aimed at the rich, the AMT has more and more applicability to middle class employees and cuts into some of the incentive value of the ISO. Some in industry and the Congress would like to eliminate the AMT altogether for ISOs.

The possible application of the AMT to the gain element of

an ISO exercise adds greater tax return planning complexity and may be an excellent reason to do some tax planning before year end with your tax advisor.

Table 6.1 Income Tax Consequences of NSOs and ISOs for Employee and Company

EVENT	NSO	ISO
• Grant of stock option at today's price for purchase (exercise) at future date.	• No taxable event to employee. • No wage deduction to company.	• No taxable event to employee. • No wage deduction to company.
• Purchase (exercise of option) of stock at grant price.	• Employee taxable event. • Federal and state tax withholding required on gain element (difference between stock's current market value and grant price). • Company has wage deduction equal to gain of employee.	• No taxable event to employee with a possible exception where the alternative minimum tax applies. • No wage deduction to company.
• Sale of stock within 1 year of exercise and 2 years from date of grant.	• Short term capital gain treatment to employee on additional gain. • No tax withholding. • No additional wage deduction to company.	• Ordinary income tax treatment to employee on gain. • Generally no federal and state tax withholding required. • Company has wage deduction equal to gain.
• Sale of stock after 1 year from date of exercise and 2 years from date of grant.	• Capital gain to employee on stock gain over and above earlier gain reported. See ISO gain section for more details about capital gains rules.	• Capital gain to employee.* • No wage deduction to company.

* *Taxed at a 28% federal rate on sales between 1 year and less than 18 months from date of exercise and 20% on sales of 18 months and over from date of exercise.*

It may be helpful in understanding the above chart if a hypothetical stock option example is used. Assume Company granted Employee an option to buy 100 shares of Company stock at the price of $10 a share on January 1, 1997 with the purchase (exercise of the option) of the shares being permitted on January 1, 1998 and the sale of the shares permitted any time thereafter. Assume the share price rises to $20 a share on January 1, 1998, to $30 a share on June 1, 1998 and to $40 a share on January 10, 1999.

Table 6.2 Tax Consequences of Exercise of Stock Option: Hypothetical Real Life Example

EVENT	NSO	ISO
• Grant of 100 shares at $10 per share, on January 1,1997.	• No taxable event to employee. • No wage deduction to company.	• No taxable event to employee. • No wage deduction to company.
• Purchase (exercise of option) of shares at $10 per share when market price is $20 per share, on January 1, 1998.	• Employee has taxable gain of $10 per share at ordinary income tax rates. • Federal and state tax withholding required. • Company has wage deduction equal to employee's gain of $10 per share.	• No taxable event to employee except for possibility of the alternative minimum tax. • No wage deduction to company.
• Sale of 100 shares on June 1, 1998 at $30 per share.	• Excess gain taxable at short term capital gains rate to employee of $10 per share. • No tax withholding. • No company wage deduction.	• Employee has taxable income at ordinary rates on $20 per share. • Generally, no federal and state withholding required. • Company has wage deduction equal to employee's gain of $20 per share.
• In the alternative, sale of 100 shares on January 10, 1999 at $40 per share.	• Taxable gain to employee of $20 per share at long term capital gains rate. • No tax withholding. • No company wage deduction.	• Employee has taxable gain at long-term capital gains federal rate of 28% on $30 per share.* • No tax withholding required. • No company wage deduction.

* If employee sold on August 1, 1999, the $30 gain will be taxed at a 20% federal rate.

Using Cafeteria Plans

Cafeteria plans represent a concept that employees want choices in the plans they are offered. Some may want medical reimbursement coverage. Younger employees may want child care benefits. Other employees may want dependent care coverage. More and more companies are tailoring their benefits packages to accommodate these diverse interests.

Two of the more common employee benefit plans are the medical reimbursement plan and the child care or dependent care plans. The employee, under U.S. tax laws, may contribute funds through payroll withholding on a pre-tax basis to these plans to pay for plan expenditures.

Medical Reimbursement Plans

These plans provide secondary medical reimbursement for those medical costs not covered by the employee's company health insurance plan. The reimbursement is limited by how many dollars the employee commits during an open enrollment period at the beginning of each plan year and then has withheld through payroll withholding. If an employee pays into the plan more than he needs to cover his medical expenses, he will lose the extra amount.

Those expenses covered by medical reimbursement plans include amounts spent that were not covered by the company plan, such as the initial deductibles. Also covered are eye glasses, prescription drugs, and dentistry not ordinarily covered by the company plan.

The employee gets to use pre-tax dollars to pay for them. The public policy issue here is that sometimes Congress looks at elimination of this benefit in its efforts to balance the budget.

Child Care or Dependent Care Plans

These plans are similar to medical reimbursement plans where employee pre-tax earnings may be put aside for such care. These plans may also be vulnerable to the budget-cutting ax.

Using Employee Tuition Reimbursement Plans: *Section 127*

One of the most important issues for companies is the training of their workforce. There is a severe shortage of well-trained people, especially in the high tech sector, forcing companies to bring in foreign workers for some jobs. One solution to the job shortage is to encourage more workers to take undergraduate and graduate courses to improve their job skills.

College training is very expensive, and costs continue to rise at a steep rate. Many employees, especially the younger ones, find continuing education prohibitively expensive. Many companies step up to help their employees pay for their continuing education. Support of employees' graduate and undergraduate training is necessary if these companies want to stay on the leading edge in the marketplace.

In the absence of some exception, tuition reimbursement paid by the employer is a taxable benefit to employees, similar to wages, and should be reported as W-2 income to them. The employer will have the legal obligation to withhold from the amount of the reimbursement federal and state income taxes, social security, and unemployment taxes.

One exception to W-2 income inclusion and employer withholding is when the training benefits the employer directly. If the employee can prove that the tuition reimbursement is related to maintaining or improving job skills and not to acquiring a college degree making him eligible for a new line of work, then no W-2 inclusion or withholding is necessary. Some companies will reimburse the employee only in these circumstances.

A second exception relates to a provision of the Internal Revenue Code (Section 127) which provides for tuition reimbursement, up to $5,250 per year, for undergraduate courses with no W-2 inclusion or employee withholding. To qualify, the employer must have a written plan that applies to all employees.

Section 127 has been in the law for many years, but its

duration has always been limited. Typically Congress extends Section 127 for only one-two years at a time. After this period Section 127 must be extended again by Congress.

This issue of extension is annoying to employers and employees. Sometimes Section 127 will lapse for several months. Employers then need to address the issue of W-2 inclusion and withholding. Such uncertainty hurts employee morale. Until now, however, Congress has always extended Section 127 retroactively to the time it last lapsed.

For some companies, Section 127 is too limited. These companies feel it should also apply to graduate level courses.

Prior to July 1, 1996, graduate courses were covered under Section 127. Then, because of federal budgetary constraints, graduate courses were eliminated, again causing employee disenchantment. Companies must now treat graduate level course reimbursements as income to the employee and withhold taxes.

The business community is pushing to extend Section 127's reach to graduate studies. The cost to the government is not cheap. Their estimate is that the cost will be about $1 billion dollars over a five year period. They say this cost will have to be made up with tax increases in other areas.

States are also looking at proposals to exempt employee tuition reimbursement. California, for example, has introduced a bill to include graduate studies.

Proponents of the Section 127 extension to graduate studies have the support of the Clinton administration and many in the Congress and state legislatures. Universities also are strong supporters.

The time to lobby for the extension to graduate studies is now. The business community is making the extension one of its highest education priorities. Employees also will be well-served to voice their interest in letters, telephone calls, faxes, and personal meetings with legislators.

This is an issue companies and employees can win!

Summary

The business culture rewards risk-taking. Compensation plans tied to profits and stock performance are typical. Bonus plans are usually tied to profitability. Stock options are tied to stock performance.

Retirement plans are typically tied to company's performance as well. Profit-sharing and 401(k) plans are the most prevalent.

Cafeteria plans allow employees to choose benefits packages best serving their needs.

Employee education is very important. Companies often reimburse their employees for undergraduate and graduate college courses. Current law permits reimbursement for undergraduate courses to be tax-free to the employee. Graduate courses are, however, taxable. With a concerted lobbying effort, the business community should be able to pass a law change to cover graduate courses.

SPANISH PESETA

CHAPTER SEVEN

WORKPLACE ISSUES

KEY TOPICS

> *Complying with the Affirmative Action Laws*

> *Avoiding Sexual Harassment Suits*

> *Avoiding Wrongful Termination Suits*

> *Assuring a Qualified Workforce: Immigration and H-1B Visas*

> *Utilizing Quality Improvement Teams*

> *Lobbying for Flex-Time*

> *Improving the Workers' Compensation System*

> *Promoting Good Ergonomics for Your Employees*

WORKPLACE ISSUES

Policies governing the workplace have become significant issues for companies. Such issues include affirmative action, sexual harassment, wrongful termination suits, immigration, flexible work shifts, employee-management quality circles, state workers' compensation programs, and ergonomics.

Complying with the Affirmative Action Laws

Affirmative action relates to programs designed by employers to help women and minorities obtain better employment opportunities as compensation for past wrongs or as a way to promote certain societal goals, such as reducing social injustice.

Affirmative action programs must not be based on quotas (which are illegal), but on voluntary good faith efforts toward achieving a workforce that reflects the diverse race, color, age, sex, or religion of society in the employer's locality.

In the debate over the success of affirmative action programs, supporters point out that affirmative action has changed the face of small business by helping many minority-owned firms get their start. Moreover, they say that affirmative action promotes a workforce that mirrors the community. While it is the morally right thing to do, it is also good business because it helps companies expand their markets into minority communities. They conclude that even more needs to be done to improve the well-being of women and minorities.

Other observers claim affirmative action programs have gone too far. A 1978 Supreme Court case found a situation of reverse discrimination (i.e., discrimination against the majority group—white males) existed and held quotas illegal in a medical school's admission policy. A similar result was found in a 1996 Texas law school admissions case. Detractors also argue that affirmative action programs are expensive, allegedly costing the nation about $20 billion annually— money that could otherwise be used to improve industry's health and competitiveness.

The policy debate has accelerated from the tug-of-war of strengthening versus weakening affirmative action to a struggle between those who would continue to implement affirmative action programs and those who seek an outright repeal. The merits of affirmative action programs will continue to be challenged in the next few years, but it seems clear the pendulum is beginning to swing toward less arbitrary programs.

Avoiding Sexual Harassment Suits

Harassment on the basis of sex is a violation of Title VII of federal law. Unwelcome sexual advances, requests for sexual favors, and other verbal and physical conduct of a sexual nature constitute sexual harassment when submission to such conduct:

> Is a term or condition of an individual's employment

> Is used as a basis of employment decisions affecting such individual

> Unreasonably interferes with the individual's work performance

> Creates a hostile working environment

Where sexual harassment exists, the employer could be found liable for damages to the employee. Between 1991 and 1997, sexual harassment charges doubled—from 8,000 to 16,000— and monetary settlements leaped from $7 million to $50 million. In 1998, Mitsubishi paid a record $34 million settlement.

Sexual harassment may include sexual slurs, threats, derogatory comments, unwelcome jokes, sexual advances, requests for sexual favors, uninvited touching, or sexually-related comments. All of these may be lumped under the category of a "hostile working environment."

A hostile working environment may not fit the legal definition of "hostile" when the conduct is not sufficiently severe and pervasive as to alter the conditions of the employee's employment. Trivial or merely annoying conduct will not be sufficient cause for a lawsuit. Hypersensitive employees may not be eligible for relief. Single, isolated incidents normally will not be considered. The conduct will be evaluated from an objective viewpoint of a reasonable man or woman facing the same conditions. The test for harassment is the victim's perspective, not community standards or stereotypes of acceptable behavior.

More case law and new federal and state laws will add definition to what is or is not sexual harassment. Employers need to be aware of new trends and continue to educate their workforce about the problems that can result from complaints of sexual harassment. Employers should also consider the purchase of employment practices liability insurance. Only when sexual harassment no longer exists will employers be free from lawsuits and payment of damages.

Avoiding Wrongful Termination Suits

U.S. employment law is undergoing a legal revolution that tends to favor the rights of employees over employers n employee termination cases. Such cases, called wrongful termination suits, are costly to defend.

Normally tried before a jury, awards have typically favored the employee and have often exceeded $500,000. This has forced many companies to hire from temporary services agencies or to outsource entire departments. After shareholder lawsuits, wrongful termination suits are one of the most common lawsuit types.

These awards have become common despite legal provisions in some states conceptually limiting them. For example, some states, including California, have a doctrine which permits an employer to terminate an employee at will (i.e., either for cause or without cause). But the employment-at-will doctrine has been narrowed by numerous federal laws and state exceptions involving public policy reasons, implied contracts and promises by employers.

Federal law exceptions to the employment-at-will doctrine include dismissal for the following reasons:

> Union affiliation

> Personal disability

> Sexual discrimination

> Race, sex, religion, or age

> National origin

State public policy exceptions grant relief to an employee where the employer has terminated the employee in violation of state public policy. Examples of terminations involving state public policy include the following types of cases:

> Refusal to commit an unlawful act (e.g., committing perjury or violating antitrust laws)

> Performance of an important public obligation (e.g., serving on a jury or engaging in whistle blowing)

> The exercise of a legal right or privilege (e.g., making a workers' compensation claim or refusing to take an illegal polygraph test)

> Unlawful discrimination

Constructive termination is a more prevalent state issue as plaintiffs and courts begin to understand the outer limits of wrongful termination. An example of constructive termination might be a situation where sexual harassment of an employee causes the working environment to be so unpleasant that the employee is forced to quit.

A state court law, often known as the implied covenant of good faith and fair dealing, involves a termination not done in good faith and not amounting to fair dealing. For example, some state courts have held that an employer may be sued for fraud when it recruits an employee with knowingly false promises, such as lifetime employment.

Jury awards to employees have involved compensatory damages for back wages, back benefits, emotional distress, reinstatement, and sometimes punitive damages. Because insurers historically have denied coverage and defense of employers for wrongful termination suits, employers have handled the costs of defending such claims directly. Lacking insurance coverage, employers tended to settle these cases out of court. Just in the past year, insurance has now become available that has coverage at more affordable prices.

In light of recent trends, employers should review their hiring practices. Look carefully at recruiter training programs, employee handbooks, policy manuals, and performance evaluation procedures. Avoid costly court proceedings by requiring new employees to agree to arbitration or mediation as an alternative. Consider the purchase of an employment practice liability insurance policy. Above all else, consult with and follow the advice of employment lawyers and other specialists in this field.

By supporting legislation that more clearly defines what is or is not wrongful termination, and by pushing for caps on damage awards, employers may be able to limit their liability in wrongful termination suits.

Assuring a Qualified Workforce:
Immigration and H-1B Visas

Immigration reform has been a hot topic for many years. Some reform proposals could have a major adverse impact on U.S. multinational companies.

U.S. industry has supported measures to stop illegals at the border, but objects to proposals to reduce the number of legal immigrants holding professional degrees. Any reduction in such professionally skilled immigrants would pose an acute problem. Representatives in the high technology industry claim a high-end worker shortage in the hundreds of thousands of workers.

Some of the labor unions dispute this claim. They say there are plenty of homegrown workers. Industry just has to give them training. By bringing in foreign workers, industry can pay them less and keep down the wages of U.S. workers in a tight labor market.

Recent immigration bills dealing with immigrants with professional skills relate to the H-1B visa. Under this category as many as 65,000 skilled foreign workers may enter the U.S. every year on temporary visas valid for up to six years. The occupations covered by the category include computer programmers, engineers, architects, doctors and college professors.

To date, such visas have permitted U.S. multinationals to place immigrant professionals in jobs in the U.S. where there is a shortage of qualified U.S. workers. An important example of this occurs in the electronics companies where foreign engineers and software designers help ameliorate a severe shortage.

Largely because of demand from high tech companies, the 65,000 cap was reached in May, 1998, causing the Immigration and Naturalization Service to stop issuing visas until October 1, 1998, when a new 65,000 visa limit takes effect with the start of the U.S. government's fiscal 1999.

Some H-1B bills have proposed a reduction of the number of professional immigrants. They require that immigrants have foreign work experience (companies could not hire them out of U.S. universities), and that they be paid a wage greater than the prevailing U.S. wage for such work. More recent bills have proposed increasing the current visa cap of 65,000 foreign workers to a number closer to 100,000 workers.

The U.S. multinational community has worked hard to keep intact many H-1B provisions of the current law and to increase the cap. This industry effort will continue for many years as debate goes on as to whether there is or is not a worker shortage.

Support of the H-1B issue is a good example of a way the business community can cooperate to change provisions in laws which are harmful to their business. By becoming politically active as a group, they have helped assure a larger, well-qualified pool of technical talent.

Utilizing Quality
Improvement Teams

Employee-management quality circle teams set up to boost productivity, quality, and efficiency may be in violation of the 1935 National Labor Relations Act.

Such teams, common in companies which are not unionized, could be prohibited by a provision in the Act which prevents companies from setting up unions controlled by management. This provision prohibits managers and employees at non-unionized companies from meeting to establish "conditions of work." Especially impacted is the high tech sector.

The National Labor Relations Board has sued several companies, including DuPont, Polaroid, and the San Jose Mercury News over this provision.

Organized labor opposes quality improvement teams and views them as an attempt to skirt unionism and to erode the balance between labor and management. The U.S. Labor Department also opposes them.

Industry has introduced bills in the U.S. Congress to permit quality improvement teams, such as the "Teamwork for Employee and Management (TEAM) Act." They have good support in the Congress but have been opposed by President Clinton. Proponents of these bills have not been able to gather the necessary two-thirds votes to override the President's veto threat.

Quality improvement teams represent an important issue for the non-unionized business community. Company quality and efficiency are at stake. Resources of time and money will need to be put toward this issue.

Lobbying for Flex-Time

Today's employers are striving to adjust employee work shifts to match more closely the needs of both employer and employee. This is often accomplished through the flex-time concept, usually consisting of longer work days and a shorter work week.

Some employee groups are arguing for flex-time rules as a way to spend more time with their families and do personal errands during normal business hours. Various other groups cite flex-time rules as an opportunity to contain child and elder care costs and reduce traffic congestion and air pollution from cars. Lower employee absenteeism may be still another benefit.

Flex-time concepts often can run afoul of overtime laws. Federal and state law require overtime pay for work over 40 hours a week. A few states require overtime pay after an 8-hour work day even if the employee does not work a 40-hour work week.

Employers assert overtime pay after an 8-hour day hurts their ability to compete against other states and foreign competitors who do not even have rules as strict as the federal 40-hour week.

Labor unions have been the primary defenders of overtime after an 8-hour work day, arguing that employers will use the flex-time concept to force employees to take time off and thus avoid paying overtime. But as labor unions have been losing membership and public approval in recent years, the pendulum may be shifting sufficiently to permit more flex-time, especially in the case of the 8-hour day overtime rules.

SAUDI ARABIAN RIYAL

Improving the
Workers' Compensation System

Workers' compensation issues have been front burner in many states in the past few years. Fraudulent claims, easy-to-file stress claims, high costs and low benefits have highlighted dissatisfaction by the business community.

Compensation issues now being addressed include the following:

> Reducing workers' compensation insurance rates

> Decreasing medical costs

> Limiting the number of medical evaluations

> Requiring a greater emphasis on workplace safety

> Limiting the costs of vocational rehabilitation

> Reducing the number of stress claims

> Streamlining adjudication procedures

> Increasing penalties for fraud

> Increasing benefits to workers

The business community will want to continue to watch workers' compensation costs and benefits, and to participate in public policy efforts to improve the system.

Promoting Good Ergonomics
for Your Employees

Ergonomics has been an issue for industry since the early 1980s. Ergonomics is defined as the science that seeks to adapt work or working conditions to the worker. Ergonomics is usually associated with repetitive motion injuries, eye strain from the use of video display terminals, and the position of chairs relative to the desk.

Federal and state regulatory bodies, responding to Occupational Safety and Health Act (OSHA) rules, have attempted to set standards of ergonomics for employers. Local communities have also attempted to set such standards, presenting companies with a confusing patchwork of requirements.

In Silicon Valley, for example, the standards are set by federal and state regulations. In San Francisco, a separate standard has been adopted and applies in those instances where it does not contradict federal and state standards.

Are these standards going too far?

Unions complain that the standards do not go far enough and should also address back injury problems, psychological stress, and additional rest breaks.

Business trade associations assert many of the standards go too far and are not based on scientific evidence showing that the standards actually help the health and safety of workers. Employers claim that the standards are also expensive to implement.

In dealing with ergonomics standards, the business community has been advancing the concept of the degree of risk versus the cost to business. At the very least, so the argument goes, the business community should have the maximum degree of flexibility in satisfying such standards.

Worker safety issues versus company cost-effectiveness will be with us for a long time. The solution should be a win-win situation. Employees should be given the tools and procedures to help them stay healthy while companies have a workforce that is able to stay on the job.

Summary

Affirmative action programs are being challenged for giving job opportunities based on gender and the color of one's skin rather than on merit.

Sexual harassment is a high profile issue in the U.S. that is now appearing in Asia and Europe.

Wrongful termination issues have generally favored the employee. Employers are well-advised to adopt policies that reduce exposure to wrongful termination lawsuits.

Immigration rules pose a problem for the business community as a shortage of qualified U.S. applicants forces companies to look overseas to fill critical job openings.

While the union movement is on the decline, unions still exert influence in such areas as immigration of foreign skilled workers, worker safety and ergonomics, the setting of flexible work shifts, and the ability of management and employees to form quality control and work-efficiency teams.

UGANDAN SHILLING

STATE AND LOCAL ISSUES

KEY TOPICS

> *Helping Ease the Traffic Gridlock for Your Employees*

> *Hiring and Developing an Educated Workforce*

> *Supporting Affordable Housing for Your Employees*

> *Lowering Your Utility Costs*

> *Cutting Through the Permitting Bureaucracy*

> *Reducing Health, Safety and Environmental Dangers*

STATE AND LOCAL ISSUES

This chapter covers some of the most important state issue areas impacting the business community. Included are discussions of transportation, education, housing, utility rates, the permit process, and health, safety and environmental issues. Not included in this chapter is the area of taxation. I have devoted all of Chapter Nine to state and local tax issues because of their critical cash flow and bottom line impact.

Helping Ease the Traffic Gridlock for Your Employees

With the rapid growth of business centers around the U.S., communities face the challenge of creating the infrastructure necessary to accommodate increasing numbers of employees. This includes housing needs, quality education, and a transportation system to get them to and from work.

Congested roads, lack of adequate public transportation, and air pollution from car tailpipe emissions are the most important transportation problems confronting communities.

Silicon Valley is a good model from which to learn. Springing up almost overnight from fruit orchards and vegetable farms, the roads in this highly populated 40- by 60-mile area were outdated. A great demand for improvements began in the early 1960s, and before long, new roads and freeways were built. But they were inadequate just as soon as they were finished. The roads became clogged and there was minimal public transportation to ease the pressure. With congested highways came the additional problem of smog. Unfortunately, a 1978 initiative, Proposition 13, limited the taxes that could be raised by cities and counties in their efforts to upgrade the highway system.

In 1984, Silicon Valley responded, primarily through the incredible efforts of a local business group called the Santa Clara County Manufacturing Group (now the Silicon Valley Manufacturing Group), which lobbied and raised money for Measure A, a half-percent local sales tax rate initiative.

Measure A won by a majority vote, which can be attributed to the Manufacturing Group's get-out-the-vote campaign. With the half-percent sales tax rate increase, the local transportation authority began to build new freeways, widen existing freeways, and commence construction of an area-wide rapid transit system. All three of these projects have greatly reduced traffic congestion.

With smog and pollution from cars and trucks, the state legislature set out to clean up the air. In 1989, the legislature passed a new tax credit law to encourage ride sharing. Special fast lanes for vehicles with two or more riders (high-occupancy vehicle [HOV] lanes) were installed on the major freeways. Cities spent money on smart traffic lights to smooth out the traffic. Smog certificates were required of drivers to curb tailpipe emissions.

Recent events have put increased stress on the Silicon Valley high technology community. A successor to Measure A was determined in 1995 by the California Supreme Court to be in violation of Proposition 62, which, the court held, requires a two-thirds majority vote to pass special taxes dedicated to specific purposes like transportation. Measure A and its successor had a majority, but not a two-thirds majority. In 1996, industry drafted Measures A and B in such a way to require a simple majority vote. Both measures passed. These measures added $1.2 billion for transportation projects.

In other developments, the state legislature eased a burden placed on Silicon Valley companies when it abolished trip reduction mandates on large employers. Trip reduction mandates required companies to encourage riders to use commute alternatives and reduce the number of driving trips they make alone in the area. This was a reasonable piece of legislation since the air in Silicon Valley is currently one of the cleanest of any major metropolitan area.

Silicon Valley companies have not always waited for legislation to tell them how to improve air quality. After a recent study showed that most auto pollution came from older cars, companies in the area formed a voluntary partnership and spent considerable amounts of money to purchase these cars and get them off the roads forever.

Still other ways Silicon Valley has responded include allowing employees to have flexible working hours in order to avoid the commute rush and, in some cases, to telecommute some work days.

Silicon Valley has been a good model for other rapidly expanding communities. The manufacturers took the initiative to work with city and county governments to maintain adequate roads and clean air without inhibiting the economic vitality of the area.

Hiring and Developing an Educated Workforce

The business community actively supports education. It has been generous in sponsoring scholarships and vocational training for its workers. Some manufacturers have provided computers for schools; others have helped build needed school facilities.

This generosity arises in great measure from self-interest. Companies need the brightest and best-educated engineers, they need manufacturing floor workers who have the skills necessary to operate the most technologically advanced equipment, and they need workers who have learned the basic reading, writing, and mathematics skills in kindergarten to 12th grade (K-12 programs). It is not uncommon for companies to allocate a percentage of their pre-tax profits for charitable endeavors and to direct most of that support toward educational initiatives.

Companies invest heavily in programs that teach vocational skills. Too many students graduate from high schools without the skills necessary to advance to the work force. Employer-sponsored community college vocational training provides one answer to this problem. Another response is in-house training. Companies spend billions of dollars annually on in-house vocational training.

Most large companies pay for undergraduate and graduate courses that lead to college degrees and improve job skills of their employees.

Companies help education in still other ways. Through adopt-a-school programs, they provide company specialists to teach classes, such as Junior Achievement economics, and other business courses. They also provide curricular enrichment activities and support services for K-12 grade students.

Supporting Affordable Housing for Your Employees

For the business community, adequate housing represents a major ingredient of a happy workforce. Housing also represents one of the keys to lowering air pollution from autos.

In many business centers, industry has grown so fast that there is an insatiable demand for local housing. This leads to rising home prices, as supply cannot keep up with demand.

With higher prices, many in the workforce find they cannot afford nearby housing and they are forced to live farther away from their employers. Workers with families often need two cars as it is increasingly difficult for one family member to drop off the other at work and then go on to another job or run errands. This increase in the number of cars on the roads leads to congestion.

Some cities, such as Sunnyvale, Mountain View, and Santa Clara, in the Silicon Valley, have begun to allow a limited number of higher density, affordable high rise housing complexes to be built near the industrial parks. They are accessible by much of the workforce, they shorten the commute, and they reduce pollution.

In Austin, Texas, employers have helped to build more

affordable housing through a federal program called the Low Income Housing Credit. Employers receive a certain percentage of the housing cost as a tax credit against their federal income tax liabilities.

More effort is still required. Companies may wish to participate in government-industry first buyer programs which help employees make down payments on home purchases. Companies may also wish to work with city planners to increase the supply of affordable housing by accelerating the review time of multi-family residential developments. This might entail more coordination among city housing departments and a one stop permit process.

The availability of affordable housing near the business centers will be critical to companies and the communities in the future. Employees are happier without the stress of long commutes; the community is cleaner with less air pollution from cars; and companies will benefit by having more alert workers who have not put in hours of tiring commuting.

Affordable housing availability represents one of the major reasons companies choose one plant site location over another.

Lowering Your
Utility Costs

An important issue for companies is the price they pay for electricity, water, waste disposal and telephone services. This issue is particularly important in plant site discussions.

Utility costs vary greatly around the country. California and states in the Northeast are known, for example, to have much higher electricity rates than areas such as the Pacific Northwest. California and the Northeast states, such as Massachusetts, Rhode Island and New Hampshire, are responding by deregulating their power industries in the hopes of reducing electricity costs. This should be good news for companies in these areas.

Water and waste disposal rates vary from community to

community. Some communities are changing vendors in antic-ipation of lower rates and better service. You should compare the rates in your community with other communities to deter-mine if your suppliers are charging you a competitive price.

Lastly, with the coming of telecommunications deregulation, fees you pay should begin to drop, thereby adding a benefit to your bottom line.

Cutting Through the Permitting Bureaucracy

Many companies, especially in the high tech fields, have com-plained over the years about how long it can take to obtain permits to build a manufacturing facility. For them, with prod-uct life cycles as short as six months, it is vital that permits be approved as rapidly as possible.

California is a good case study of this problem. Delays were particularly acute in California in the 1980s and early 1990s, where high tech companies complained that even simple tenant improvements or equipment permits could take between 30-90 days to process. By that time, the market opportunity for their companies could be lost.

Because of problems like these, in addition to expensive land and construction costs, a workforce shortage and high wages, many Silicon Valley companies began an exodus to other states. Texas, Oregon, Arizona and New Mexico landed many plant site expansions of prestigious Silicon Valley companies.

What many of these other states offered was easy permitting. Rumor has it that a Silicon Valley company visited Austin to look at a possible plant site location and was met at the airport by the Texas governor and Austin mayor with all of the neces-sary permits in hand. It is not surprising that Austin is one of the leading high tech centers today.

To counter the flight of Silicon Valley companies to other states, cities in Silicon Valley teamed up with industry representatives to explore ways to streamline the permitting process. They set out to attack cumbersome fire, safety and environmental building codes with the goal of reducing the permit time. Over a period of several years, they have identified and corrected many time delay procedures in the permitting process. Chief among these improvements are the following:

> Coordinating city staff into one building for easy access

> Permitting payment of fees with credit cards or monthly bills

> Establishing one-stop permit shopping with a turnaround of one day

> Appointing a single point of contact—to shepherd the permits required by a company

> Allowing electronic filing of permit applications

> Adopting uniform fire, safety and environmental building codes for cities in the area

> Developing uniform rules and interpretations for the building codes

> Working closely together with a company in the pre-permit stage to resolve issues before building commences

> Publishing simple, smaller, easy-to-read permit handbooks

> Increasing industry—city communication from the beginning to the end of the building process

As a result of Silicon Valley's efforts to streamline the permitting process, companies are now reporting a marked improvement in the speed and efficiency of getting permits issued. This translates into cost savings and projects completed on schedule.

Facilities managers in the semiconductor industry in Silicon Valley, for example, report that cities such as Sunnyvale now compete on an equal footing with Austin for permitting efficiency and speed.

For preferred projects, it is not uncommon now for state government executives to work with city government executives to cut through the red tape and assist a company in becoming operational in the shortest time possible. I am aware of one major city setting up a special team composed of the city manager and key department heads to clear the major permitting hurdles and help the company through the permitting bureaucracy in the shortest time possible.

Because of this emphasis on speedier permitting in California and other states, U.S. industry has become more competitive with its foreign competition. Those companies still experiencing permitting delays will be well-served to form city-industry work groups to study what states like California and Texas have done and adopt many of their permitting reforms.

Reducing Health, Safety and Environmental Dangers

An area of costs that might appear to add greatly to your operating expenses is expenditures to improve the health, safety and environmental working conditions of your employees and the community. In the short term, your costs will be impacted. In the long run, however, you ought to be able to avoid some costs that otherwise would apply had you not taken your short-term steps.

Some of the key health issues you need to address include how working conditions impact the health of your employees and the offering of company benefits aimed at preventing employee health problems before they become serious. You will want to have your employees in well-lighted areas to cut down on eye strain, encourage frequent breaks from working at computers, and provide ergonomically proper chairs and desks to avoid repetitive motion hand injuries and back strain.

Offering simple in-house medical exams for measuring blood pressure and eyesight and giving instruction on stress reduction, good diet and exercise habits may provide early warnings to your employees about current and future health problems, thus saving your company future cost outlays for expensive medical procedures.

In the safety area, companies need to take steps to insure safe working conditions. This can be as simple as offering safety courses. Companies may also need to consider automating some functions that involve risks, such as using robots to handle dangerous chemicals.

Working conditions, generally safe for most workers, may not be safe for the handicapped, women who are pregnant, and some older workers. Alternative working environments should be made available to them. Awareness of the danger and taking the appropriate safety steps will prevent many workers' compensation claims and costly lawsuits.

Chapter Two contains commentaries on environmental dangers involving air, water and leaking underground storage tanks. Containing costs in these areas might involve taking advantage of pollution control tax credits and tax-exempt revenue bonds for pollution abatement equipment.

Many companies have become more proactive in reducing hazardous waste through such means as substituting nonhazardous materials in the manufacturing process and recycling waste water. Greenhouse gases reduction is another priority for many companies.

From my personal experience, companies have become environmentally more sensitive today. Most want to comply to the maximum extent possible. And yet, poorly drafted and unreasonable regulations, interpreted and enforced by over-zealous regulators, may add greatly to your compliance costs.

In your next site selection analysis, you should discern those cities and states with efficient and effective environmental laws from those cities and states affording no better protection but snaring a company in a ribbon of red tape.

Summary

Transportation issues are important in areas experiencing rapid population growth. As more plants are built and more people are hired, there are greater demands on the transportation system. Road expansions and alternative forms of transportation will be important issues to resolve.

Education is a problem for the business community. There are not enough qualified people to fill the job openings. Companies often must provide the education themselves. Industry is studying ways to improve the education system.

Business communities need to address the issues of affordable housing and utility rates, local permitting bureaucracy, and health, safety and environmental issues.

ENGLISH POUND

ARGENTINEAN PESO

PHILIPPINE PESO

STATE AND LOCAL TAXATION

KEY TOPICS

> *Lowering Your State Income Tax Bill*

> *Working with the Unitary Tax System*

> *Minimizing Your Company's Property Taxes*

> *Using the Sales and Use Tax Rules to Your Advantage*

> *Containing Other Local Taxes and Fees*

> *Taxing the Internet—Good or Bad Idea?*

STATE AND LOCAL TAXATION

State taxes impact a company's ability to be competitive. This chapter discusses state income tax issues, the famous (and controversial) unitary tax area, the area of property taxation and major sales and use tax issues. Included are comparative state charts in the areas of income, property, and sales and use taxes. Also discussed are issues relating to local taxes and fees and taxation of Internet transactions.

Lowering Your State Income Tax Bill

As cash-starved state governments increase their tax rates, broaden their tax base, and reach out to tax transactions beyond their borders, companies feel the impact.

For many companies, state income taxes pose a greater financial burden than federal income taxes. Until a few years ago, the business community sat back and accepted the increasing tax burdens from their states.

More recently, industry has aggressively fought tax increases and has lobbied hard to create state business climates more in line with competitor states and countries. For companies in states without an income tax, such as Texas and Washington, this discussion has limited importance. But for others, new pro-business tax laws have helped relieve onerous taxation.

Perhaps the greatest tax benefit to manufacturing companies has been the change in many states from a three-factor to a four-factor income apportionment formula. California, New York, and Oregon are examples of states now using a four-factor formula, which often results in lower taxes for companies that are building plants and hiring people within these states. (States use the four-factor formula as an economic development tool.) State income apportionment is discussed further in the next topic.

An incentive for economic development has been the research and experimentation (R&E) tax credit. California and Massachusetts lead other states in their support of R&E spending. The manner in which the credit is calculated in California is addressed in the federal R&E tax credit discussion in Chapter Four.

Another tax incentive used by some states is the investment tax credit. California, for example, provides a 6% investment tax credit to medium and large manufacturers. This credit is very important to many companies. New manufacturers receive a sales tax exemption. Refer to the sales and use tax section in this chapter for more information on the California investment tax credit and sales tax exemption provisions.

Working with the Unitary Tax System

In unitary taxation, corporate business income is figured in proportion to a company's business presence in a state (apportionment). Income tax then applies to the apportioned income.

Unitary refers to the dominant type of income for a business. In the electronics industry, for example, all income relating to the sales of electronic products is normally classified as unitary. Income for an electronics company in areas unrelated to the electronics business, such as income from a separate real estate or timber business, might appropriately be classified as non-unitary income. It would be apportioned separately from electronics income.

A company's business presence in a state typically is measured by its sales, property, and wages in the state compared as a percentage to its sales, property, and wages in all of the states, and, in some cases, in the world.

In states like California, which apply the worldwide apportionment method, unitary taxation has been a public policy sore point. Several countries, including Japan, Canada, the United Kingdom, and the Netherlands, have challenged the unitary method as taxing the worldwide income of multinationals from these countries when such companies may have had no income (perhaps even losses) in the state. In the past few years, California has modified its law to address the foreign company concerns.

The method of apportionment will vary from state to state. Many states use an equally weighted three-factor formula test of sales, property and wages. An increasing number of states, such as California, use a four-factor formula, with sales representing double the normal factor rating, and property and wages being a single-factor rating each. To the extent that a company does business in several states, each using a different type of apportionment formula, some double state taxation could result. Examples of the three-factor and four-factor double weighted sales formulas are illustrated below.

Example:

Parent-subsidiary group of companies in similar lines of business who collectively earn pre-tax profits of $1,000 and have business activities in California and elsewhere as follows:

TURKISH LIRA

Assume a 10% California tax rate.

1. THREE-FACTOR FORMULA (Old California Method):

TAX BASE	WORLDWIDE	CALIFORNIA	CALIFORNIA PERCENT
Sales	$50,000	$10,000	20%
Property	$10,000	$5,000	50%
Wages	$25,000	$12,500	50%

Add sales, property and wage percentages for California and divide by 3.

$$20\% + 50\% + 50\% = 120\%$$
$$120\% \div 3 = 40\%$$

Multiply pre-tax profits of $1,000 by .40 = $400.

California tax on California apportioned profits:
Multiply $400 by .10 tax rate = $40.00

2. NEW CALIFORNIA METHOD (Double Weighted Sales Factor Formula)
Using the same numbers as above, the calculation would be as follows:

TAX BASE	CALIFORNIA PERCENT	
Sales	20%	
Sales	20%	= 140%
Property	50%	
Wages	50%	

Add percentages and divide by 4
$$140\% \div 4 = 35\%$$
Multiply $1,000 by .35 = $350

California tax on California apportioned income: Multiply $350 by .10 tax rate = $35.

As you can see from a comparison of the three- and four-factor formulas, the four-factor double-weighted sales formula results in a considerable tax savings where a company has a small sales base but a heavy property and wage base in a state. The four-factor formula will not be as good as the three-factor formula, however, when the sales percentage in the state is greater than the property and wage percentages. Texas' single (sales) factor formula provides even greater incentive to locate jobs and manufacturing plants in that state. Massachusetts law is phasing in to a single sales factor formula as a way to out-compete states with double weighted sales factor formulas.

Table 9.1 State Corporate Income Taxes: A Comparison of Major States (as of March 31, 1998)

STATE	HIGHEST CORPORATE TAX (approximate)	APPORTIONMENT FORMULAS [1] RATE (%)
Arizona	9.00	4
California	8.84	4
Colorado	5.00	3
Connecticut	9.50	4
Florida	5.50	4
Georgia	6.00	3
Idaho	8.00	4
Illinois	4.80	4
Maine	8.93	4
Maryland	7.00	4
Massachusetts	9.50	1 [2]
Michigan	2.30	4 [3]
Minnesota	9.80	4 [4]
New Hampshire	7.00	4

Table 9.1 State Corporate Income Taxes:
A Comparison of Major States (as of March 31, 1998) Cont.

STATE	HIGHEST CORPORATE TAX (approximate)	APPORTIONMENT FORMULAS [1] RATE (%)
New Jersey	9.00	4
New Mexico	7.60	4
New York	9.00	4
North Carolina	7.25	4
Ohio	8.90	4
Oregon	6.60	4
Pennsylvania	9.99	3
Rhode Island	9.00	3
South Dakota	0.00	No
Texas	0.00	1 [5]
Utah	5.00	3
Vermont	9.75	3
Virginia	6.00	3
Washington	0.00	No

1) Formulas designated as follows:
Single weighted (sales only):1
Normal 3-factor formula of sales, property, and wages: 3
Normal 3-factor formula of sales, property, and wages where sales have double weighting: 4

2) Massachusetts is gradually going to single weighted sales factor after 1999

3) Michigan is gradually going to 90% sales factor after 1999, with property 5% and payroll 5%

4) Minnesota weights sales 70% and property and wages 15% each

5) Texas has a franchise tax that uses a single-weighted sales factor formula

Minimizing Your Company's Property Taxes

As companies have grown, so have the cost of their plants and equipment, administrative buildings, and in-house computer infrastructure. A state's property tax burden could well be an important issue for a company considering locating in that state.

Property tax rates vary from state to state. California, with the 1978 passage of Proposition 13, for example, has a low tax rate of 1% of fair market value, with valuation increases capped at 2% a year. Texas, another well-known high technology state, has tax rates in the range of 2% to 3% applied to fair market value.

For a capital-intensive company in the semiconductor industry, construction of a two billion dollar state-of-the-art wafer fabrication facility could result in a first year property tax burden of $20 million in California ($2 billion x .01) or $50 million in Texas ($2 billion x .025). These are not insignificant amounts.

The manner in which a state provides special incentives or realistic depreciation schedules for equipment may go a long way to alleviate a high property tax rate.

Some states grant property tax abatements of up to 10 years on new plants and equipment. Oregon's method of taxing only the first $100 million of investment has recently attracted several new semiconductor facilities with investment costs well in excess of $100 million for each facility.

Other states have made an effort to accelerate the depreciation rates of equipment as an incentive. In Arizona and Texas, for example, property tax depreciation schedules have been accelerated for computer and semiconductor equipment. And at the city and county level, some states have further discretion to abate city and county taxes. Cities in Silicon Valley have recently been providing some small abatements in this area.

Behind the push to offer property tax abatements and incentives by state and local governments is the competition not only from other states, but also from foreign countries. Many high technology companies expanding overseas, for example, have negotiated large incentive packages which include free land, large cash grants for building and equipment purchases, tax holidays, and grants for employee training.

It is not surprising that competition for high tech companies is so fierce. These industries represent some of the crown jewels of U.S. manufacturing. Countries in Europe and Asia see high technology strength (and perhaps dominance) as a way to improve the standard of living of their citizens. Some states have been aggressive in their plant site bidding for the same reasons. Those states currently leading in high tech employment and investment will not want to rest on their laurels. They can still do more to retain or attract new high tech investment.

In the capital-intensive sector of high tech, which includes biotech, semiconductor equipment, and semiconductor wafer fabrication facilities, states have to appreciate the huge investments these companies need to make, perhaps ten times the dollar levels of the 1980s, and ask themselves if it is fair to fully tax the investments of these companies.

If the state compares an investment in a semiconductor wafer fabrication facility of the same land size and building square footage as a Wal-Mart store, should one be taxed more than the other? A state-of-the-art semiconductor facility costs $2 billion today. A state-of-the-art Wal-Mart costs perhaps $20 million. With the number of cars and customers at a Wal-Mart, arguably it requires more of the services property taxes are supposed to cover than does a semiconductor facility. And yet, in California the semiconductor facility would have a property tax burden of $20 million on a $2 billion investment, while the Wal-Mart would pay $200,000 on a $20 million investment. It is not surprising semiconductor and other high tech companies are looking at other locations.

The area of property tax burdens will be an extremely important public policy issue over the next several years. Those companies trying to maximize their bottom lines will want to look carefully at this area. Aside from the need to have property tax expertise within your company, many law and accounting firms offer specialists in this area also.

For further information, Table 9.2 sets out, by major state, property tax rates, inventory (freeport) exemptions, and, in a general way, property tax abatement policies.

The chart is limited to personal property tax policies, since personal property is generally the major issue area for the business community.

Real property taxes exist for all of the states listed and, in some cases such as in Oregon, it is possible to obtain an abatement from real property taxes as well.

VIETNAMESE DONG

Table 9.2 Personal Property Taxes:
A Comparison of Major States (As of March 31, 1998)

STATE	INVENTORY EXEMPTION YES/NO	PERSONAL PROPERTY EXEMPT?	AVERAGE PROPERTY TAX RATE	ABATEMENTS & OTHER INCENTIVES
Arizona	Yes	No	3.50%	Yes
California	Yes	No	1.10%	Yes
Colorado	Yes	No	3%	No
Connecticut	Yes	No	3.9%	No
Florida	Yes	No	2.50%	No
Georgia	No	No	4.15%	No
Idaho	Yes	No	1.75%	No
Illinois	Yes	Yes	No personal property tax	No Tax
Maine	Yes	No	2%	Yes
Maryland	No	No	3%	No
Massachusetts	No	Yes[4]	No personal property tax	No Tax
Michigan	Yes	No	4.70%	No
Minnesota	Yes	Yes	No personal property tax	No Tax
New Hampshire	Yes	Yes	No personal property tax	No Tax
New Jersey	Yes	Yes[5]	No personal property tax	No Tax
New Mexico	Yes[1]	No	3.00%	Yes
New York	Yes	Yes	No personal property tax	No Tax

Table 9.2 Personal Property Taxes:
A Comparison of Major States (As of March 31, 1998) Cont.

STATE	INVENTORY EXEMPTION YES/NO	PERSONAL PROPERTY EXEMPT?	AVERAGE PROPERTY TAX RATE	ABATEMENTS & OTHER INCENTIVES
North Carolina	Yes	No	2.15%	No
Ohio	No	No	8.10%	No
Oregon	Yes	No	1.50%	Yes
Pennsylvania	Yes	Yes[6]	No personal property tax	No Tax
Rhode Island	No	No	4.28%	No
South Dakota	Yes	Yes	No personal property tax	No Tax
Texas	No[2]	No	1.75%	Yes
Utah	Yes	No	1.75%	Yes
Vermont	Yes[3]	No	2.58%	No
Virginia	Yes	No	3.00%	Yes
Washington	Yes	No	1.50%	No

[1] NM exempts inventories with a few exceptions

[2] Freeport exemption available if goods exported within 180 days of purchase

[3] Towns have the option of allowing an exemption

[4] Corporate personal property is exempt from local property taxes but is included in the property measure of the corporate excise (income) tax

[5] Fixtures are taxable in NJ as of January 1996

[6] Personal property taxes are imposed only on intangible personal property

Using the Sales and
Use Tax Rules to Your Advantage

As states push for more revenues, the area of sales and use taxes becomes fertile ground for rate increases and a broadening tax base.

The issue of "nexus" (a taxpayer's connection to a state) is constantly being tested in the areas of out-of-state mail order houses and electronic transfers as states seek more sales and use taxes.

There is discussion at the federal and state levels about the wisdom of taxing new types of Internet transactions. Government sees it as a source of new funds. The business community argues that such taxation of the Internet will slow down its use and great potential. This issue will be on-going for many years to come.

Sales taxes generally apply to the purchase of tangible personal property, such as equipment, within a state. In transactions where no sales tax applies, a complementary tax, called a use tax, may sometimes apply.

For example, when a Texas company buys equipment in Texas, pays Texas sales tax on it, and then ships the equipment to its plant in California, California can impose its use tax. If the California tax rate is 8-1/2% and the Texas rate is 8%, double taxation will not result because California law allows a credit for the Texas tax, thus leaving a remaining California tax of .5%. Hereafter, I will use the term sales tax to refer to sales and use taxes.

Industry's public policy agenda in this area has been to limit the scope of the tax either by definition or exemptions.

The issue of definition could arise, for example, in the case of R&D contracts or the sale of software coupled with hardware. The general rule is that sales taxes apply only to tangible personal property. The taxing authorities have often tried to treat R&D (a service) as bundled with a prototype (tangible property) or software (intangible) as bundled with hardware (tangible property). Once bundled, the taxing authorities try to apply the sales tax to both. The dispute as to what is tangible and what is not will continue for many years.

The issue of exemptions often arises when tangible property is consumed in the manufacturing process or when equipment is purchased for use in manufacturing. It is common for exemptions to be issued in these situations, although the form of exemption will vary from state to state.

For example, in 1994 California provided a modified exemption for the purchase of equipment. The equipment was still taxable for sales tax purposes, but in certain situations the buyer could claim a 6% investment tax credit against its income tax liability. Certain new California manufacturers received a sales tax exemption of 6% for 1994 and 5% thereafter.

Most states, however, provide an outright sales tax exemption for manufacturing supplies and equipment.

For those California companies reporting losses, the California investment tax credit law poses a problem. The credit will not currently apply. Those companies with a string of loss years could lose the credit altogether due to limited credit carry forward provisions. Since they cannot claim the credit currently, profitable California companies have a competitive advantage over them. And yet, such loss companies are doing exactly what the California legislature asked of them—buy new equipment and hire more people. This inequity should be addressed by the California legislature by providing a sales tax exemption or similar relief, such as a refundable credit, for equipment purchases.

With the high cost of supplies and equipment used in the manufacturing process, companies will be prudent to pay close attention to the public policy issues regarding sales and use taxes. Many companies employ in-house sales and use tax experts. Again, some law and accounting firms offer specialization in this area.

The following chart shows sales tax rates and exemptions, if any, for supplies and equipment used in the manufacturing process.

TABLE 9.3 Sales And Use Taxes:
A Comparison of Major States (as of March 31,1998)

STATE	TAX RATE PERCENTAGE (Includes avg. city & county percentages)	EXEMPTION FOR CONSUMABLE SUPPLIES USED IN THE MANUFACTURING PROCESS	EXEMPTION FOR MANUFACTURING EQUIPMENT
Arizona	5.00-8.00	No	Exempt
California	7.25-8.50	No	Partial Exemption or Mfg. Investment Credit
Colorado	3.00-8.50	Yes	Exempt if Purchase is > $500
Connecticut	6.00	Yes	Exempt
Florida	6.00-7.50	No	Generally Exempt
Georgia	5.00-7.00	No	Exempt
Idaho	5.00-7.00	Yes	Exempt
Illinois	6.25-8.75	Yes	Exempt
Maine	6.00	Yes	Exempt
Maryland	5.00	No	Exempt
Massachusetts	5.00	Yes	Exempt
Michigan	6.00	Yes	Exempt
Minnesota	6.50-7.50	Yes	Taxable at Reduced Rate
New Hampshire	No Tax	N/A	N/A
New Jersey	6.00	Yes	Exempt
New Mexico	5.125-6.875	Partial	Exempt[1]
New York	4.00-8.50	Yes	Exempt
North Carolina	6.00	No	Taxable at Reduced Rate

TABLE 9.3 Sales And Use Taxes:
A Comparison of Major States (as of March 31,1998) Cont.

STATE	TAX RATE PERCENTAGE (Includes avg. city & county percentages)	EXEMPTION FOR CONSUMABLE SUPPLIES USED IN THE MANUFACTURING PROCESS	EXEMPTION FOR MANUFACTURING EQUIPMENT
Ohio	5.50-7.00	Yes	Exempt
Oregon	No Tax	N/A	N/A
Pennsylvania	7.00	Yes	Exempt
Rhode Island	7.00	Yes	Exempt
South Dakota	4.00-6.00	Yes	Taxable
Texas	6.25-8.25	Yes	Exempt
Utah	5.75-7.75	No	Exempt as of 7/98 Previously taxable
Vermont	5.00	Yes	Exempt
Virginia	4.50	Yes	Exempt
Washington	7.50-8.60	No	Exempt

[1] New Mexico exempts equipment purchased with industrial revenue bonds. This is a common way to purchase equipment. Otherwise, the equipment is taxable.

Containing Other
Local Taxes and Fees

City and county taxes and fees can add considerably to a manufacturing company's cost structure. Whether you currently are located in a city or considering the city for your next plant site location, you need to carefully evaluate these costs. Manufacturing companies may be able to reduce their operating costs by one to two percent a year by selecting one city location over another.

Of course, companies choose city locations for many reasons. Nontax reasons, such as land and building costs, wage rates, infrastructure, availability of a qualified labor pool, and proximity to your customers will usually outweigh tax considerations. But once you have narrowed down your city choices based on nontax reasons, local taxes and fees will be important in making your final city selection.

City and county revenue sources include property taxes, sales and use taxes, city income, payroll and gross receipts taxes, business license fees, permitting fees, utility user's taxes, and redevelopment fees. A typical city revenue stream, by percentage, is as follows:

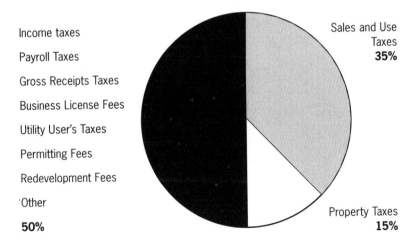

Income taxes
Payroll Taxes
Gross Receipts Taxes
Business License Fees
Utility User's Taxes
Permitting Fees
Redevelopment Fees
Other
50%

Sales and Use Taxes **35%**

Property Taxes **15%**

Cities and counties face numerous operating expense pressures and are always looking for ways to enhance their revenue base to cover police, fire, roads, sanitation and other needs. Taxing various new types of Internet transactions has appeal to many localities as a way to increase their revenue stream. Taxing of the Internet is discussed later in this chapter.

Responsible companies will want to support the revenue needs of their city location to make sure they have adequate city services and a good community environment for their employees. Companies also, however, will want to keep an eye on city costs.

Some local costs are easier to determine than others. If you look at city locations in California, for example, the easier costs to determine would be property and sales and use taxes.

In California, real property taxes are based on one percent of your manufacturing plant's land and building costs, as increased each year by an inflation factor of two percent. The property tax rate is the same throughout the state, though local direct assessment and bond measure taxes differ.

The sales and use tax rate varies with each county. If you were looking at possible plant site locations in the counties of Los Angeles, San Francisco, Santa Clara (Silicon Valley), San Diego, Monterey and Sacramento, you would note the following rate differences:

SALES TAX RATE DIFFERENCES BY SELECTED COUNTIES

Los Angeles	San Francisco	Santa Clara	San Diego	Monterey	Sacramento
8.25%	8.50%	8.25%	7.75%	7.25%	7.75%

The above rates consist of a state rate of 6%, a combined city and/or county rate of 1.25%, and transactions (transit/transportation) taxes which range between .125% and 1.25%, depending upon whether the sale occurs within a special taxing district.

A major revenue source for some cities is an income tax.

Several major cities, such as New York, have an income tax that is a percent of the state income tax. This can add a substantial extra cost to your operations.

Other cities, such as San Francisco, impose a payroll tax and a gross receipts tax. The payroll tax is a set dollar amount per San Francisco employee while the gross receipts tax applies to those sales and other transactions attributable to San Francisco. The employer pays a tax based on the higher of the two taxes. Many employers in San Francisco claim the tax makes the city uncompetitive.

Less obvious types of taxes consist of business license fees, permitting fees, utility users taxes and redevelopment fees. Such fees and taxes will often be paid by an operating division of the company and miss the scrutiny of the company's tax department. These taxes are harder to research in a plant site selection analysis and many times are overlooked entirely.

The city of San Jose represents a good example of the magnitude of these taxes for companies. For 1996, San Jose had a revenue budget of about $300 million dollars, of which about $8 million came from business license fees and about $92 million from permitting, utility user and redevelopment fees and taxes. Industry paid a substantial percentage of this tax burden.

Suffice it to say that local taxes and fees should not be overlooked. They need to be taken into account when doing a plant site selection analysis. They also need to be understood and watched by companies already in the locality to make sure tax increases are proper and not an unjustifiable cost burden on industry.

Watching and containing these costs perhaps best can be done by the facilities department in a company in conjunction with the tax department and senior management. Hiring an outside consultant to advise you in this area might also be worthwhile. Sometimes former city managers and city financial officers make excellent consultant choices.

Taxing the Internet
—Good or Bad Idea?

The Internet is rapidly changing the way the world communicates and does business. More people are using the Internet every day. Traditional ways of communicating and buying products and services are yielding to new Internet media. Government is eyeing these mediums as possible new revenue sources.

What are some types of Internet transactions that might be subject to taxation?

> E-mail

> Buying of goods through the Internet

> E-Trade for the buying and selling of stocks

> Advertising and providing information (web pages, travel information, current news)

> Internet access charges

Should these and similar transactions be subject to taxation? Is taxing transactions over the Internet a good or bad idea?

One might think of various tax issues with the Internet. At the federal level, there is the issue of taxes on communications which would apply to your telephone calls and faxes but not to E-mail. E-mail is made possible through telephone equipment— so to the extent you connect via the telephone, E-mail is taxed as communications.

There may be foreign tax reporting issues if a foreign country claims Internet transactions as having an income tax source in that country, and the U.S. says such a transaction has a source in the U.S. This could lead to double taxation (See Chapter Five).

For the most part, however, attention to the taxation of Internet transactions has focused on the state and local issues, and typically involves doing business ("nexus") questions in a state or local area and whether income, sales or use, or some other method of taxation could apply.

Opponents of taxation say unwise taxing of Internet transactions will kill the Internet in its infancy. They claim unreasonable compliance burdens and new taxation will discourage Internet use. Industry has lobbied federal and state legislative bodies to place moratoriums on imposing new types of taxes on the Internet.

Congress has responded with bills such as the Cox-Wyden "Internet Tax Freedom Act," which would place a moratorium on the ability of state and local governments to impose new taxes on the Internet or on-line services for a period of up to six years. Some states, such as California, have also introduced moratorium bills. These moratoriums do not prevent local jurisdictions from imposing and collecting existing taxes on Internet-related transactions.

Industry's rationale for not taxing new Internet transactions can be summarized by three concerns:

> The administrative burden for taxpayers

> The potential for new onerous taxes

> The potential for double taxation of taxpayers—first in another state and then in their home state

These concerns are illustrated by the following example. If you order products from an out-of-state mail order house, such as L.L. Bean in Maine, should L.L. Bean be required to collect sales tax in your home state if L.L. Bean has no business operations there?

If you order from L.L. Bean over the Internet, is the transaction any different than calling L.L. Bean's 800 toll-free number?

In either case, opponents of taxation would say it is administratively too burdensome for L.L. Bean to know all of the estimated 30,000 state and local sales tax rates for every place it ships product. The obligation should not be on L.L. Bean, they claim, but on the buyer who has the legal responsibility to report a use tax liability to his home state.

Of course, we all know that most buyers would not so

report. Thus, we have a compliance problem and a loss of revenues for states and localities.

The U.S. Supreme Court has agreed with L.L. Bean and probably would apply the same reasoning to the above Internet sale. This decision puts the burden on the buyer to pay a use tax and file a use tax return in the buyer's home state.

State and local governments claim there should be ways to tax some Internet transactions. Most are concerned about a loss of sales tax revenues. For many localities, sales tax revenues may make up 25% to 50% of their revenue base, money necessary to fund police, fire, roads and other functions. They cite studies showing that major department stores in their locality will contract or close down in favor of locating in tax friendly areas and engaging in business over the Internet from out-of-state locations.

One local government estimate states that each 5% shift of nationwide sales from store front to cyberspace leaves 110 million square feet of retail selling space vacant—the equivalent of about 2,500 Price Clubs or 100 regional malls that would no longer be needed. This will, they believe, result in a loss of sales tax revenues for the locality due to the rapid acceleration of transactions over the Internet, as well as a loss of jobs, particularly the low wage and welfare-to-work jobs.

State and local governments have the support of many state governors, mayors and the National League of Cities.

The solution to the above problem lies in a methodology that provides adequate tax revenues for the state or locality while not imposing unfair administrative burdens on out-of-state sellers. Responsible companies do not want to see state and local governments drained of the resources necessary to provide essential services.

One possible approach is to adopt a federal sales tax at one rate or at different rates depending on the product and require out-of-state sellers to collect it.

For simplicity and to avoid disputes over the types of items that should be taxed, only physical goods, such as books, clothes, etc. would be taxed. Services, software downloading and other types of contentious items would not be subject to tax collection.

These tax revenues could then be allocated among the various states and localities based on the out-of-state seller's sales destinations. For example, an allocation within a state might be 60% to the state and 40% to the cities and counties.

To be administratively easy, the seller would file only one tax return in its home state showing the out-of-state sales destinations, the tax collected, and the tax payment to the home state. The home state then would have the obligation to apportion such federal tax to all of the relevant state and local jurisdictions around the country.

This approach would not catch all sellers (e.g., from foreign countries) or satisfy all states and localities (e.g., the federal rate might be lower than the state rate), but it would be a start in the collection of "new found" money that otherwise would not have gone to the state and locality.

With the Internet, we are in new tax waters. Creative thinking is essential to permit state and local governments to fund their essential services while not imposing unfair tax and unworkable administrative burdens on taxpayers and killing the promise of the Internet.

New types of taxes on Internet transactions today are a bad idea. Federal and state moratorium laws for a period of years are necessary until thoughtful, fair solutions can be determined. While it may take years to arrive at comprehensive solutions, piecemeal legislation, such as the federal sales tax discussed above, has merit and should be explored. As the old saying goes, "A good plan today is better than a perfect plan tomorrow."

This issue will continue for many years. It is important for companies to voice their concerns and work closely with state and local governments for a win-win solution.

Summary

Some states are competing for your business through lower income tax rates or special incentives.

Companies are discovering that sales and property taxes impact pretax profits and can be burdensome.

A state's property taxes can be the factor that determines whether or not it is able to attract a major investment.

Local taxes and fees vary greatly from locality to locality. Their cost impact should not be overlooked.

Taxes relating to Internet transactions represent the latest contentious issue between government and industry.

ITALIAN LIRA

ECONOMIC DEVELOPMENT AND PLANT SITE SELECTION

KEY TOPICS

> *Understanding State Economic Development Programs*

> *Selecting Your Next Capital-Intensive Domestic Plant Site Location*

   Table 10.1 Factors to Consider in Selecting a Plant Site (Worksheet)

> *A Tale of Two Cities: Comparing Austin, Texas and Sunnyvale, California for a Capital-Intensive High Technology Investment*

   Table 10.2 A Comparison of Austin and Sunnyvale for a Capital-Intensive High Technology Investment

> *Selecting Your Next Software Company Site Location*

   Table 10.3 A Comparison of Eleven Software Company Site Locations in the U.S.

> *Selecting Your Next International Plant Site Location*

   Table 10.4 A Comparison of Twenty International Plant Site Locations

ECONOMIC DEVELOPMENT AND PLANT SITE SELECTION

This chapter helps us understand the relationship of taxes and other cost issues in the context of state economic development programs and plant site selection criteria. Included is a plant site comparison of two important high technology cities: Austin, Texas and Sunnyvale, California. Also considered are a software company site selection analysis for eleven U.S. software centers and an international plant site selection analysis of twenty major countries for a capital-intensive investment.

Understanding State Economic Development Programs

States are in fierce competition to attract manufacturing projects. What are some of the enticements states offer to attract such projects?

Non-tax enticements:

> Fast permitting

> Can-do attitude

> Easy-to-apply environmental laws

> Workplace and tort laws that are neutral or favor employer

> Free land

> Training grants

> Low workers' compensation rates and high benefits

> Good transportation system

> Good worker training programs

> Abundant labor supply

> Low water, gas and electricity rates

> Low worker wages

> Good educational system

> Affordable employee housing

Tax enticements:

> Low or zero state income tax rate

> Single-factor (sales) or double-weighted apportionable sales factors based on sales, property, and wages

> R&D tax credit

> Investment tax credit

> Sales tax exemptions for equipment, utilities and tangible personal property consumed in manufacturing and R&D activities

> Property tax abatements

> Low property tax rates

> Fast property tax write-off of equipment

In evaluating non-tax and tax enticements, leading-edge companies often put together a team of company experts to add input to the process. Such teams might include technical experts, strategic planning and marketing experts, and various other experts from human resources, finance, tax, and legal departments. The leader of the team is usually a top executive of the company.

Selecting Your Next Capital-Intensive Domestic Plant Site Location

What state will be the winner in your next plant site selection process? Using a state-of-the-art semiconductor wafer fabrication facility as a model, with land and building costs of $200 million and manufacturing equipment costs of $800 million, what non-tax and tax factors will be key to your ultimate decision?

I have used a ranking system of one to five, with one being the worst and five being the best. Some items can be weighted extra (perhaps double or triple) to show their increased importance. After you give the items a numerical ranking, add them to see which site has the highest score.

Table 10.1 Factors to Consider in Selecting a Domestic Plant Site Location (Worksheet)

NON-TAX CONSIDERATIONS	5	4	3	2	1
Land costs					
Building costs					
materials					
labor					
Available water and waste water					
Available gas and electricity					
Trained workers					
electrical engineers					
technicians					
other					
Colleges					
electrical engineering schools					
vocational schools					
K-12 school quality					
Proximity to international airport					
Ability to locate in foreign trade zone					
Receptivity of community					
Receptivity of state					
Availability of cultural events					
Availability of recreation					
Labor costs					
electrical engineers					
technicians					
other					
Availability of affordable housing					
Permitting time					
Proximity to customers					

Table 10.1 Factors to Consider in Selecting a Domestic Plant Site Location (Worksheet) cont.

NON-TAX CONSIDERATIONS	5	4	3	2	1
Workers' compensation rates and benefits					
Environmental law bureaucracy					
Natural disaster sensitivity					
Supplier network					
Proximity to top management					
Litigation friendliness					
Transportation system					
TAX CONSIDERATIONS					
State income tax rate					
Availability of net operating loss provision					
R&D credit					
Investment tax credit					
Unitary method					
3-factor formula					
double-weighted sales formula					
single formula—sales					
Sales tax rate					
Sales tax exemption for supplies used in manufacturing process					
Sales tax exemption for manufacturing equipment					
Property tax rate					
Property tax exemptions					
Property tax equipment depreciation					
Property tax abatements					
Local taxes and fees					

A Tale of Two Cities:
Comparing Austin, Texas and Sunnyvale, California for a Capital-Intensive High Technology Investment

Austin, the high technology center in Texas, has an edge over cities in Silicon Valley, such as Sunnyvale, but that edge has been decreasing as California state and local governments have taken steps to make California high technology investments more competitive. If Austin were compared to the lower cost area of Sacramento, California, an up-and-coming high technology center, the two conceivably might be even.

The following is my assessment of how Austin and Sunnyvale compare in a few selected areas.

CHILEAN PESO

Table 10.2 A Comparison of Austin and Sunnyvale for a Capital-Intensive High Technology Investment

SUBJECT	AUSTIN	SUNNYVALE
Land Costs	Edge	
Building Costs	Edge	
Water and Waste Water		Even
Utility Costs		Even
Trained Workers Pool		
Engineers		Even
Technicians and Other		Even
Colleges		Edge
Receptivity of Community		Even
Receptivity of State		Edge
Labor Costs	Edge	
Affordable Employee Housing	Edge	
Permitting Time		Even
Workers' Compensation Rates and Benefits		Even
Environmental Law Bureaucracy		Edge
Supplier Network		Even
Litigation Friendliness	Edge	
Transportation System		Edge
State Income Taxation [1]		Even
Sales Taxation	Edge	
Property Taxation [2]		Edge

[1] Texas has no income tax, but it does have a franchise tax that is generally not too burdensome. California, with its double-weighted sales factor apportionment formula, the R&D credit, and the investment tax credit, comes close to the low income tax burden of Texas.

[2] Austin has a higher average property tax rate (2.75% vs. 1.10%). Texas communities will sometimes provide a property tax exemption. Austin is today rarely providing such exemptions.

Selecting Your Next
Software Company Site Location

If you are a startup software company or an existing software company planning to expand your operations in the U.S., where should your new site be?

Joint Venture: Silicon Valley Network commissioned the accounting firm of Ernst and Young to do such a plant site study.

Ernst and Young examined the costs incurred by a "prototypical" software company with first year revenues of $50 million. The study assumed an expansion of 400 employees and 100,000 square feet of space in Silicon Valley and ten other U.S. technology centers.

The study, completed in 1997, considered how business expenses in these eleven areas over a five year period would affect the company's profitability.

Expenses in the study involved:

> Salaries and wages

> Employer taxes

> Contract and outside services

> Equipment purchases

> Materials and supplies

> Real estate and utilities

> Insurance and personal property taxes

The rankings of these software centers, from the lowest cost region to the highest cost region, were as follows:

Table 10.3 A Comparison of Eleven Software Company Site Locations in the U.S.

RANK	REGION
1	Salt Lake City
2	Seattle
3	Portland
4	Austin
5	Denver
6	Dallas
7	Raleigh
8	Atlanta
9	Phoenix
10	Boston
11	Silicon Valley

1997 Ernst and Young study

Selecting Your Next
International Plant Site Location

Some of the same considerations you addressed in your U.S. plant site decision-making analysis apply to selecting an international plant site. In addition, there are many other considerations to be taken into account.

> Will English be spoken?

> What is the level of bureaucracy in the country?

> Will bribery (corruption) be a factor?

> What are the political and financial risks of the country?

> Is intellectual property protected?

> What are the various tax and other incentives?

At Table 10.4, I give my perspective for some of these issues for twenty foreign country locations.

My country opinions may not always coincide with yours, but I hope you will find this analysis useful.

Because of limitations of space, I have omitted some important countries, such as Israel, Greece, Spain, Chile, South Africa, Australia, and New Zealand. One could easily argue that these countries should have been in the study.

For the analysis, I use a five point rating system with five being the best and one being the worst. I consider 35 issue areas. One might think of many more. As some issue areas will be more important than others, you could argue that they should be double- or triple-weighted.

I have assumed a capital-intensive manufacturing facility, a high-level of trained workers, expatriates, a need to protect intellectual property, and the need for some software design capability. My views come from a U.S. company's perspective.

I have included some large population countries because of their future economic significance and the opportunities for companies investing there. India intrigues me because it will pass China in population by 2050, is an English speaking country, and yet remains an unknown for many companies.

China and Turkey impress me for their interest in becoming economically more successful and for their huge consumer markets. Russia has great potential and is one of my favorites to become a future powerhouse. Indonesia appears to be a sleeping giant and will awake someday.

In this analysis, Canada came out on top, followed by Ireland, Hong Kong, England and Singapore. In the worst category are Indonesia, India, Russia, and China.

The total scores from Table 10.4 are as follows:

COUNTRY	RANK
Canada	143
Ireland	139
Hong Kong	137
England	136
Singapore	136
France	131
Germany	130
Taiwan	129
Italy	128
Malaysia	120
Japan	117
Turkey	117
Thailand	114
Mexico	113
Korea	109
Brazil	108
China	99
Russia	99
India	96
Indonesia	89

Table 10.4 A Comparison of Twenty International Plant Site Locations

OPERATING CONSIDERATIONS	Ireland	England	France	Germany	Italy	Russia	Japan	China	Hong Kong	Taiwan	Singapore	India	Indonesia	Korea	Thailand	Malaysia	Canada	Mexico	Brazil	Turkey
English Language Barriers	5	5	3	3	3	2	2	1	4	2	4	4	1	2	2	4	5	3	3	3
Bureaucracy	5	5	4	5	3	1	3	1	5	4	5	1	1	3	3	3	5	3	3	3
Corrupt Conditions (illegal under U.S. laws)	5	5	4	4	3	1	4	2	4	3	3	1	1	3	2	3	5	2	2	2
Intellectual Property Protection	5	5	5	5	5	2	4	2	4	4	4	2	2	3	3	3	5	3	3	3
Industrial Espionage Risk	5	5	5	5	5	2	4	2	4	4	3	2	2	3	3	3	5	3	3	4
Political Risk	5	5	5	5	5	2	4	2	4	4	4	2	1	3	3	3	5	3	3	3
Land Costs	3	1	1	1	1	5	1	5	1	3	3	5	5	3	4	4	2	3	3	3
Building Costs	2	1	1	1	1	5	1	5	3	3	3	5	5	3	4	4	4	2	3	4
Permitting Simplicity	5	5	5	5	5	1	4	2	4	4	4	1	1	3	3	3	5	3	3	3
Labor Costs Engineers	2	1	1	1	1	5	1	5	4	4	4	5	5	3	5	5	2	4	3	4
Technicians	2	1	1	1	1	5	1	5	4	4	4	5	5	3	5	5	2	4	3	3
Other	2	1	1	1	1	5	1	5	4	4	4	5	5	3	5	5	2	4	3	3
Labor Law	4	4	3	3	3	3	5	5	5	5	5	3	4	5	5	4	4	3	4	3
Supplier Network	4	5	5	5	5	2	4	2	4	4	4	2	2	3	3	3	5	3	3	3
Domestic Transportation System	4	5	5	5	5	2	4	2	4	4	4	2	2	4	3	3	5	3	3	3
Utilities Availability and Reliability	4	5	5	5	4	2	4	2	4	4	3	2	1	3	3	3	5	3	3	4
Quality of Educational System	5	5	5	5	5	4	5	3	4	4	4	3	1	3	2	3	5	3	4	3
Vocational Training Availability	5	5	5	5	5	4	4	2	4	4	3	2	2	3	3	4	5	3	3	4
International Flights/Frequency	4	5	5	5	5	2	4	2	5	4	5	2	2	3	3	3	5	3	3	4

Table 10.4 A Comparison of Twenty International Plant Site Locations (cont.)

	Ireland	England	France	Germany	Italy	Russia	Japan	China	Hong Kong	Taiwan	Singapore	India	Indonesia	Korea	Thailand	Malaysia	Canada	Mexico	Brazil	Turkey
OPERATING CONSIDERATIONS																				
Expatriate Housing and Goods Affordability	2	1	1	1	1	1	1	2	1	3	2	4	2	3	4	4	4	4	4	4
Expatriate Personal Safety Conditions	4	5	5	5	5	2	4	2	4	4	5	3	2	3	3	3	5	3	3	3
Hotels/Restaurants/Amenities	3	4	4	3	3	3	3	4	4	4	4	4	3	3	4	3	3	4	4	4
Cultural Events	4	5	5	5	5	5	5	4	4	4	4	3	3	3	2	2	4	3	3	3
Recreational Availability	4	5	5	5	5	5	5	4	4	4	4	3	3	3	2	2	4	4	3	4
Receptivity of Country to Investment	5	5	5	5	5	5	5	4	5	4	4	3	3	4	4	4	4	4	3	4
FINANCIAL CONSIDERATIONS																				
U.S. Export Control Considerations	5	5	5	5	5	2	5	1	4	4	5	3	4	5	4	4	5	5	3	5
Convertibility of Currency into U.S. Dollars	5	5	5	5	5	3	5	3	5	4	5	2	3	4	4	4	5	4	3	4
Dividend Repatriation Availability	5	5	5	5	5	3	5	3	5	4	5	3	3	4	2	4	5	4	3	4
Local Ownership Requirements	5	5	5	5	5	3	5	3	5	4	5	2	2	3	3	3	4	4	3	3
Monetary Stability	5	5	5	5	5	3	5	4	5	4	5	3	1	2	2	2	5	3	3	3
Corporate Tax Rate (income and other)	4	2	2	1	2	3	1	3	4	3	3	1	3	3	4	4	2	3	3	3
Tax and Other Investment Incentives	5	4	4	4	5	4	2	4	3	3	3	3	3	3	4	5	4	3	4	3
Tariff Rates	5	5	5	5	5	1	5	1	5	4	5	1	3	3	3	3	5	4	4	4
Expatriate Taxation	2	1	1	1	1	1	1	4	4	4	4	4	3	4	5	5	3	3	4	4

Summary

There is fierce competition for your investment at the state and international level.

Those states and countries successful in attracting companies have worked hard to create a low-cost and supportive environment.

A good education system is essential.

Tax incentives are important.

The ability to help with financing can be critical.

Fast permitting and simplicity of bureaucracy can be attractions in site selection.

FRENCH FRANC

BELGIAN FRANC

SWISS FRANC

CHAPTER ELEVEN

POLITICAL AND CHARITABLE ACTIVITIES

KEY TOPICS

> *Contributing to Politicians*

> *Setting up In-House Government Affairs Departments*

> *Giving to Charities*

> *Writing to Public Officials*

> *Meeting Public Officials*

POLITICAL AND CHARITABLE ACTIVITIES

Political giving, in-house government affairs departments, and charitable giving are important areas for the business community. This chapter discusses the benefits of becoming more politically and socially active. Also included are tips you can use when writing or meeting public officials.

Contributing to Politicians

Why political giving? The pragmatic answer is that corporate giving is a hard fact of life, like it or not. Give or fall behind.

Many sectors of the business community are woefully immature in the political giving process. The unfortunate result is that these sectors have not been as successful in their public policy programs as they could be.

So pull out your wallet and be part of the process. How do you do this?

At the federal level, corporate contributions to candidates are not allowed except when made by a federal Political Action Committee (PAC). Individual contributions are more common and legal up to certain dollar limits.

Corporate contributions may, however, generally be made to Democratic, Republican, and other official parties where the purpose of the contribution goes toward efforts to get people to vote or to fund national convention expenses. These latter contributions are sometimes referred to as "soft money" contributions.

At the state level, the rules vary greatly. In Texas, for example, corporate political giving to candidates is illegal.

In California, contributions to state and local candidates are legal. Some contributions have dollar limits.

Political contributions to politicians are necessary because running for office is very expensive. Your involvement might make the difference in getting good candidates elected.

In California, for example, the race for a U.S. Senate seat may require the winner to raise $30 million. A congressional seat may cost the winner $2-4 million. A state senate seat may cost $1-2 million and a state assembly seat $500,000 to $1 million. Winning the governor's race may require $30-60 million.

The point should be obvious. Politicians may view your industry sector as the most important industry to our country's future, but they also need to raise a lot of cash for their campaigns. As a result, their legislative platforms will reflect in great part the needs of industries and other groups (unions and trial lawyers) that have made the greatest contributions to help them get elected. Intellectual honesty and patriotism often will yield to the source of the funds.

Setting Up In-House Government Affairs Departments

Do you need to set up a government affairs department? Perhaps.

Companies with annual sales of $10-100 million may benefit from a public affairs specialist working part-time on public policy issues. If sales are over $100 million, it may be appropriate to have a full-time employee. When sales exceed $300-400 million, a full-time specialist is absolutely essential to maximize your company's sales and profits.

Old management school thinking used to concern itself with theories of designing, manufacturing, and selling. New management school thinking has added a fourth element—external affairs.

Your company may have the best designed product, the most cost efficient manufacturing, an insatiable consumer demand, and the best marketing and sales force. Yet, you may not succeed if external forces cause you to limit your international sales because of U.S. export control restrictions, or if currency rate fluctuations make your dollar-based products uncompetitive, or if foreign predatory pricing and dumping by deep pocket competitors cause you to exit your critical markets. A government affairs specialist could help you successfully address these external forces.

Tax and financial policy considerations may also adversely affect your profits. High tax rates cut into profits. Lack of appropriate tax incentives may make you uncompetitive. Or the Financial Accounting Standards Board (FASB) may attempt to require you to treat stock option exercises as an expense on your income statement, causing you a net loss and corresponding drop in earnings per share. Your stock price goes down. This drop in your stock price might come just as you are about to go to the stock market for new equity capital (a sad, but common scenario!). A government affairs specialist will help you improve your bottom line.

These examples are painfully real. Some semiconductor equipment companies were hurt a few years ago when they realized they were not eligible for a California investment tax credit on special purpose clean room buildings used in their manufacturing because of an omission of language in the legislation covering their buildings.

A number of the companies in this industry sector individually lost millions of dollars of investment tax credit. A few dollars spent for a government affairs specialist would have enabled them to be sure these buildings were eligible for the credit.

Amending the law became a priority tax issue for the trade association, Semiconductor Equipment and Materials International (SEMI), and in 1996 the industry succeeded in changing the California law to allow the credit for their clean rooms on a prospective basis.

How do you pick a government affairs representative?

In the old days you might choose someone with a degree in political science. The trend today is often to choose a more technically oriented individual, typically someone with a law degree or technical background. This move to "technocrats" has been necessitated by the complicated tax and legal issues facing industry today.

The ideal candidate will have congressional and executive branch experience in Washington, D.C., know the state legislature and local county officials, and be familiar with the European Commission and government bureaucracies in Japan and Asia. This individual should also know how to work with industry trade associations, such as the American Electronics Association and the National Association of Manufacturers.

Government affairs is a bottom line effort. But don't expect results overnight. Successful government affairs departments spend years establishing rapport and trust with elected officials and their staffs and government agency personnel. If you are patient, your government affairs department will reward your company with open and unrestricted markets, less government bureaucracy, and fewer intrusions on running your business. This will lead to tax and financial successes that will add greatly to your bottom line.

Giving to Charities

Corporate charitable giving is an important component of government affairs and the political giving process. Why give to charities? It feels good, but as a business issue, it may also have many important ramifications for your company.

The contributions you make to local charities will boost employee morale. Your employees may take more pride in working for you because of the good deeds you are doing in the community.

Making contributions to the health and welfare of the surrounding neighborhood may make you the pride of the community and perhaps gain you a more sympathetic audience with the city council the next time you need a variance for a planned expansion. Your local, state, and federal politicians will also like you for helping improve the community.

Adopt a local elementary or junior high school. Your company can help improve the quality of education that your employees' children receive if you lend resources (money, equipment, and volunteers) to train them for the information age.

It is to be hoped that these examples will whet your appetite for charitable giving. But how do you go about giving?

First, hire an employee (part or full-time depending on your size and financial ability) to set up a charitable giving program. This person should have experience and training in charitable giving and a big heart. This is not a job for Scrooges.

Next, establish a charitable giving budget. How much do you want to contribute? Do you want to contribute a yearly flat amount or an amount that varies with profitability? Many recommend the latter because it is tied to the fortunes of your company. Typically, companies contribute about 1% - 2% of pre-tax profits.

Finally, establish clear guidelines for your program which can be widely publicized. Non-profit agencies that receive your guidelines will do some self-screening which will save your staff time in the long run. You should clearly identify which kind of agencies will be ineligible to receive funds (religious and advocacy groups, for example).

Besides contributing to schools and local health and welfare charities, you could also do a matching gifts program (matching dollar for dollar gifts made by your employees) to higher education institutions and local non-profit 501(c)(3) organizations. You could also contribute your old furniture or excess inventory (e.g., computers) to local charities.

A charitable giving program, combined with a government affairs program that also contains political giving, will enhance your prospects of success in increasing your sales and bottom line profits.

Writing To
Public Officials

The most effective letter is a personal one, not a form letter. It should be concise, informed, and polite. Some specific tips:

> Type letters of not more than one page. If writing longhand, write legibly.

> In the first paragraph, state your purpose. Stick with one subject or issue. Support your position with the rest of the letter.

> If writing about a bill or regulation, cite it by name and number.

> Support your position with information about how legislation (or a regulation) may affect you and others.

> If you think legislation (or a regulation) is wrong and should be opposed, say so, indicate the likely adverse effects, and suggest a better approach.

> Ask for the official's views, but do not demand support.

Meeting
Public Officials

A meeting with a public official is an effective way to emphasize your interest in an issue or bill. Some specific tips:

> Make an appointment, state the subject to be discussed, the time needed, and identify attendees.

> Determine beforehand your spokesperson, if others are going with you, and agree on your presentation.

> If a staff person substitutes for the public official, do not be surprised or disappointed. This is a common occurrence for busy public officials.

> Know the facts related to your position. If a bill, know the number and title.

> Present the positive impact of the solutions you support and the problem it corrects.

> Discuss the negative impact of positions you oppose, and suggest, where appropriate, a different approach.

> Leave fact sheets of no more than one or two pages.

> Request favorable consideration, thank the public official and leave promptly.

Summary

The business community needs to be more proactive in its political and charitable efforts.

More attention needs to be paid to the political process. Industry needs to put up more money to get business-oriented candidates elected.

A government affairs representative could help your company increase its sales and enhance its profits.

By combining charitable giving with political giving, industry will be better able to shape public policy in its favor.

ISRAELI SHEKEL

SUMMARY

Many of the top public policy issues that impact the business community change from year to year, but those I have selected for discussion should be relevant for many years.

CEOs and their management teams will face many challenges as they guide their growing companies toward success in a worldwide market. Survival may be difficult.

The U.S. is less dominant in worldwide trade. As the dollar weakens in value, our buying power is reduced. When the dollar is strong, our exports suffer.

Although emerging competitor countries want access to the U.S. market, they may not want to give the U.S. access to theirs. Foreign dumping in the U.S. market has been a problem. There is little protection for intellectual property in most of Asia, in areas of the old Soviet Union, or in Latin America. Some of our companies face long regulatory delays in getting their products to market.

The cold war is over, but U.S. policy on export controls still presents a problem to exporters of software and hardware technology with encryption capability and to exporters of leading edge products to certain countries, such as China.

U.S. tax policy towards overseas operations of American companies remains in misguided shape, putting us at a disadvantage with our foreign competitors who enjoy superior depreciation and other cash flow generators.

Our enlightened policies toward affirmative action, sexual harassment, wrongful termination, and employee health and safety issues are among the best in the world, but the cost of these programs is an expense our foreign competitors often do not have. In the long run, the higher prices that American manufacturers are forced to place on their products may result in declining sales and a loss of U.S. jobs.

We are the most litigious country in the world, with more lawyers per capita than any other country. Frivolous securities and product liability awards have been burdensome. The high cost of litigation is yet another factor which affects our ability to remain competitive in the world market.

We have learned that the more distant we are from the seat of government, the more difficult it is to get things done. We must not overlook the local arena. Attention to state and local matters may help speed up the permitting for a new plant, reduce our utility costs, ease environmental bureaucracy, and lower our income, property, and sales tax bills.

In this competitive world, an advantageous plant site location can make the difference between survival and going out of business. Many regions of the world are fighting hard to attract your business. Possession of facts and perspective and a sharp pencil are critical in making the right location choice.

Industry has made progress in the public policy arena as more companies send their CEOs and government policy representatives to Washington, D.C. and the state capitals. We are better known and regarded today. Business and government are working together much better. Our timing is right. We have a good story to tell. Legislators want to help us. We have to be present to educate them.

INDEX

A

abatements and other incentives (personal property taxes), 182, 183-184

accelerated depreciation, 109-111, 180

Acer, 44

Advanced Micro Devices (AMD), 23

Advanced Microelectronics, 44

Advanced Pricing Arrangements (APA), 134

affirmative action, 149-151, 160, 220

Afghanistan, 65

Airbus, 130

air pollution (Clean Air Act), 69-70, 79-80, 86

Alternative Minimum Tax (AMT), 109-110, 117-119, 120-121

alternative tax systems, 109, 119-121
see also flat tax, value added tax

America Online, 98

American Airlines, 91

American Civil Liberties Union (ACLU), 95

American Electronics Association (AEA), 36, 215

Ameritech, 91, 95

anti-dumping, 2, 33, 40-42, 43-45, 54-56, 67-68
see also dumping

antitrust, 87-96
see also Clayton Act, Competition Policy, joint ventures, monopolies, Sherman Act, strategic combinations, tying arrangements

apportionment formulas, 174-179
see also formula tests—three-factor, four-factor, single-weighted sales, double-weighted sales, hybrid

Argentina, 61

Arizona, 167, 178, 180, 183, 187

arms length transfer pricing (international taxation and customs), 132

Association of Southeast Asian Nations (ASEAN), 60, 61

Austin, Texas, 197, 198, 202-203

Australia, 206

Austria, 55, 58-59

average property tax rates, 183-184

B

Baby Bells, 95

Bank of America, 23

BankAmerica, 91, 95

Belgium, 55, 59

Bell Atlantic, 65, 91

BIT, 44

block exemptions, 94

Boeing, 91, 130

Brazil 31, 61, 106, 207-209

Britain, see United Kingdom

British Airways, 91

Brussels, 55-56

business review procedure
(antitrust), 94

C

cable companies, 94-95

cafeteria plans, 139, 145, 148

California 22-23, 60, 73, 82-83, 102,
112-113, 147, 152, 163, 166-167,
169, 174-178, 180-181, 183, 185-
186, 190, 193, 197-198, 202-203,
213-214
see also Proposition 13, Proposition
62, Proposition 211

Cambodia, 65

Canada, 60, 105, 176, 207-209

capital gains, 109, 116-118, 120-121,
141-144
see individual tax rates, corporate
federal capital gains rate, dynamic
model approach, static model
approach

Caribbean Basin Initiative (CBI),
47, 65

Castle and Cooke (Dole Foods), 23

Chase Manhattan, 91

Chemical Bank, 91

child care plans, 145

Chile, 61, 206

China, 23, 28-29, 33-34, 36-37, 50,
56, 62, 64-68, 105, 206-209, 219

Chrysler, 91

class action securities suits, 72-75

classification, product (customs), 46

Clayton Act, 90

Clean Air Act, 69, 79-80

Clean Water Act, 69, 80-81

clinical testing expenses, 114, 115

Cold War, 48-50

Colorado, 178, 183, 187

comfort letters (antitrust), 94

Commerce Department, see United
States Government

Committee on Foreign Investment in
the United States (CFIUS), 62

Commonwealth of Independent States
Countries (former Soviet Union), 36

Compaq, 91

compensation (employee), see
employee compensation

compensatory damages, 76

Competition Policy (antitrust), 93

Computer Systems Policy Project
(CSPP), 50

Connecticut, 178, 183, 187

Continental, 91

constructive termination (wrongful
termination), 153

contributions
—charitable, 211-212, 215-216, 218
—political, 211-212, 215-216, 218

controlled foreign corporations, 130

Coopers & Lybrand, 91

copyright laws, 87, 97

corporate federal capital gains rate,
116

corporate federal income tax rate,
116, 124, 126, 128-129, 137

Council of Ministers (EU), 55

countervailing duties, 33, 40, 42, 56

country of origin, 46

Cox-Wyden Internet Tax Freedom
Act, 193

Cuba, 65

customs, 33-34, 45-48, 53, 56, 60,
65-68, 132-134
see classification, Caribbean Basin
Initiative, country of origin, duty
suspensions, free trade zones, manufacturing drawback, non-tariff barriers, Generalized System of

Preferences, same condition drawback, valuation

D

Daimler-Benz, 91

deep pockets (lawsuits), 76

Defense Department, see United States
Government

defined benefit plan (taxation), 141

defined contribution plan (taxation),
140

Denmark, 55, 56

dependent care plans, 145

depreciation, 110-111, 121, 180, 200,
220
—accelerated, 110-111,180

Digital Equipment Corporation, 91

Directives (EC), 55-56

Directorate(s) General (EC), 55-56
—DGI, 55
—DGIII, 56
—DGIV, 56
—DGXXI, 56

directors' and officers' insurance
policies (D&O insurance), 73-74

Domestic International Sales
Corporation (DISC), 128

double taxation, 124, 126, 138, 176,
192, 193

double-weighted sales, 176-177, 179,
201, 203

DRAMs (Dynamic Random Access
Memory), 38-39, 41, 43-44, 99-100

drawback, duty, 47
—manufacturing, 47
—same condition, 47

dumping margins (of high tech
products), 43-44
—Japan, 43
—Korea, 44
—Taiwan, 44
see also anti-dumping

Du Pont, 156

duties, see customs

duty suspensions, 47

dynamic model approach, 117
see also capital gains

E

Economic and Monetary Union, 58

economic development, 197-199

Economic Strategy Institute, 50

education, 24, 98, 146-148, 162,
164-165, 171, 197, 208, 210, 216

employee compensation, 139-142
see profit sharing, 401(k), pension
plans, stock purchase plans, stock
option plans, cafeteria plans,
medical reimbursement plans,
tuition reimbursement plans

employment practice liability
insurance, 154

encryption, 33, 50-53, 68
see also export controls

Energy Department, see United States
Government

England, see United Kingdom

environment, 34, 39, 69-70, 79-82,
84, 86, 91, 151-153, 161-162, 168-
171, 190, 198, 201, 203, 210, 220
—and business community, 79-83
—encouraging responsibility
through tax policy, 84-85
see also Clean Air Act, Clean Water
Act, leaking underground storage
tanks

EPROMs (erasable programmable
read-only memory), 38, 43

ergonomics, 150, 158-159

Ernst & Young, 91, 204, 205

euro, 33, 56-59

Eurocrats, 55

Europe 1992 (EU plan), 56

European Commission (EC), 55, 56,
59, 93-94

European Monetary Institute, 58

European Union (EU), 23, 34, 37, 46-47, 55-58, 60-61, 68, 92-94, 105
—described, 55-56
see also European Commission, member countries by name (Austria, Belgium, Britain, Denmark, Finland, France, Germany, Greece, Ireland, Italy, Luxembourg, Portugal, Spain, Sweden, The Netherlands)

Exon-Florio Act, 33-34, 61-62

export controls, 29, 33-34, 48-53, 68, 209
see also encryption

F

fabless producers (semiconductor industry), 44

facilitating payments, 63-64
see also Foreign Corrupt Practices Act)

fair use (copyright laws), 98

Fairchild Semiconductor Corporation, 62

Federal Communications Commission (FCC), 95

federal taxation, 109-148
see also alternative minimum tax, alternative tax systems, capital gains, depreciation, independent contractor vs. employee, Foreign Sales Corporation, Foreign Tax Credit, Low Income Housing Credit, orphan drug tax credit, research and experimentation tax credit, retirement plans, Sections 482 and 936, stock options

Federal Trade Commission (FTC), 91, 93-94

Financial Accounting Standards Board (FASB), 128, 214

Finland, 55, 59

flat tax, 119, 121

flex-time, 157

Florida, 178, 183, 187

Food and Drug Administration (FDA), 77-79

foreign availability (export controls), 49

Foreign Corrupt Practices Act, 33-34, 63-64

Foreign Sales Corporation (FSC), 123, 128-130, 138

Foreign Tax Credit (FTC), 123-128, 138, 192

foreign subsidiary, 130-134

formula tests, 174-179
—three-factor, 176-177, 179
—four-factor, 176-177, 179
—double-weighted sales, 176-177, 179
—single-weighted sales, 177, 179
—hybrid, 179
see also apportionment formulas, inventory exemptions, state income tax rate

401(k), 139-141, 148

France, 55, 59, 207-209

Free Trade Area of the Americas, 61

free trade zones (FTZ), 47

frivolous securities suits, 69-70, 72-73, 86

Fujitsu, 43, 62

G

GATT (General Agreement on Tariffs and Trade), 36, 45, 48, 53, 96, 99-100, 128
—panel, 100
—Rounds, 53
see also World Trade Organization (WTO), individual rounds by name (e.g. "Kennedy Round")

Generalized System of Preferences (GSP), 46, 65

Georgia (U.S.), 178, 183, 187

Germany, 55, 59, 106, 207-209

Global Information Infrastructure (GII), 51

global warming, 84

Goldstar, 44

government affairs, 211-215, 218

Greece, 55, 206

Guam, 47, 128, 135

H

H-1B visa, 149, 154-155

Hitachi, 43

Hazardous Materials Management Ordinance (HMMO), 82

Hewlett-Packard, 85

high occupancy vehicle (HOV) lanes, 163

Hong Kong, 105, 207-209

Hyatt, Gilbert, 96

hybrid formula (taxation), 179

Hyundai, 44

I

Idaho, 178, 183, 187

Illinois, 178, 183, 187

immigration, 149, 150, 154-155, 160

Immigration and Naturalization Service (INS), 155
see also H-1B visa, immigration

Incentive Stock Options (ISOs), 118, 142-144

income tax, 84, 116-118, 120, 122, 125, 128, 135-137, 141, 143-144, 146, 166, 173-175, 178-179, 188-192, 199, 201, 203

independent contractor, 109, 110, 115-116, 121

India, 28-29, 37, 50, 105, 206-209

individual capital gains rates, 116, 117

individual tax rates, 116, 120

individual validated licenses, 49

Indonesia, 28, 105, 136, 207-209

insurance, see employment practices liability insurance, directors' and officers' insurance

Intel, 92

intellectual property, 96-108
see also copyright laws, misappropriation of trade secrets, piracy, Section 337, Semiconductor Chip Protection Act, "Special 301", statutory protection periods, submarine patents

intercompany transfer pricing (international transfer pricing between related companies), 132-134

Internal Revenue Code(s), see Section(s)

Internal Revenue Service (IRS), 115, 129

international taxation, 123-138
see also controlled foreign corporations, Foreign Sales Corporation, Foreign Tax Credit (FTC), intercompany transfer pricing, Subpart F income, tax treaties, U.S. Possessions Income Tax Credit (Section 936)

International Trade Commission (ITC), 41, 42, 100

international trade, 33-68
see also countervailing duties, customs, dumping and anti-dumping, euro, European Commission, European Union, Exon-Florio Act, Foreign Corrupt Practices Act, GATT rounds, intellectual property piracy, International Trade Commission, Jackson-Vanik amendment, Most Favored Nation, NAFTA, Section 301, Section 337, "Special 301", Super 301, U.S.-Japan Semiconductor Trade Agreement, World Trade Organization, and individual countries

international transfer pricing (between related companies), 132-134

Internet, 173-74, 185, 190, 192-196 see also Cox-Wyden, double taxation

inventory exemptions, 182, 183-184

investment tax credit, 175, 186, 199, 201

Iran, 48-51, 65

Ireland, 55, 59, 105, 207-209

Israel, 65, 206

Italy, 55, 59, 105, 207-209

Iraq, 48-51, 65

J

Jackson-Vanik Amendment, 64, 65

Japan, 37, 43, 92, 99-100, 105, 110, 132-134, 136-137, 176, 207-209, 215

joint and/or several liability, 72, 76

joint ventures, 87-89
—production, 89
—research and development, 88-89
see also Microelectronics and Computer Technology Corporation (MCC), National Cooperative Research Act of 1984, National Cooperative Production Amendments of 1993, SEMATECH, strategic combinations

Justice Department, see United States Government

K

Kennedy Round (trade), 53

key recovery (encryption), 52

Korea, North, 52, 65

Korea, South, 23, 28, 39, 43-44, 51, 65, 99-100, 103, 105, 206-208

KPMG Peat Marwick, 91

L

L.L. Bean, 193

labor, 29, 65, 154-157, 189, 198, 200, 203, 208
—unions, 154
see also National Labor Relations Act, National Labor Relations Board

Labor Department, see United States Government

Laos, 65

Lawrence Livermore National Laboratory, 83

lawsuit
—exposure, 69-70, 72, 75-77, 86, 220
—abuse, 69-72, 86
see also Clean Air Act, Clean Water Act, Food and Drug Administration, frivolous securities lawsuits, leaking underground storage tanks, product liability, punitive damages

leaking underground storage tanks (LUST), 69-70, 80, 82-83, 170

Lemelson, Jerome, 96

Libya, 48, 51, 65

Lockheed Martin, 91

Los Angeles (County), 79, 190

Low Income Housing Credit, 166

Luxembourg, 55, 59

M

Maastrict Treaty, 58

Maine, 178, 183, 187, 193

Malaysia, 46, 136-137, 207-209

MAMCO Manufacturing, 62

manufacturing
—drawback, 47
—exception under Subpart F, 131
—exemptions (consumable supplies, equipment) 187-188

marking (customs), 46

Maryland, 178, 183, 187

mask works (semiconductor industry), 102

Massachusetts, 166, 175, 177-178, 183, 187

Matsushita, 43

McDonnell Douglas, 91

MCI, 91

Measure A (Santa Clara County Transportation Initiative), 163

Measure B (Santa Clara County Transportation Initiative), 163

medical reimbursement plans, 145

mergers and acquisitions, 90-92

method of apportionment, see apportionment formulas

Mexico, 60-61, 106, 207-209

Michigan, 178, 183, 187

Microelectronics and Computer Technology Corporation (MCC), 88

Microsoft, 92, 115, 130

Minnesota, 178, 183, 187

Mitsubishi, 43, 151

monopolies, 90, 94, 95

Montenegro, 65

Monterey County, California, 190

moral persuasion, 71

Most Favored Nation (MFN), 33-34, 54, 64-66, 68

Mountain View, California, 165

munitions (export controls), 52

N

National Association of Manufacturers, 215

National Cooperative Production Amendments of 1993, 89

National Cooperative Research Act of 1984, 88

National Labor Relations
—Act, 156
—Board, 156

National League of Cities, 194

National Security Agency, 52

National Semiconductor, 62

NationsBank, 91

NEC, 43

negative clearances (antitrust), 94

Netherlands (The), 55, 59

New Hampshire, 166, 178, 183, 187

New Jersey, 179, 183-184, 187

New Mexico, 179, 183-184, 187-188

New York State, 174, 179, 183, 187

New York City, 191

New Zealand, 206

nexus (taxation), 185, 192

Nonqualified Stock Options (NSOs), 142-144

non-tariff barriers, 28, 34, 48, 54

non-tax enticements, 198-199

North American Free Trade Agreement (NAFTA), 33-34, 60-61, 65, 68

North Carolina, 179, 184, 188

Northrop Grumman, 91

Northwest, 91

Nynex, 91, 95

O

Occupational Safety and Health Act (OSHA), 158

Ohio, 179, 184, 188

Oki, 43

Omnibus Trade and Competitiveness Act, 37

Oracle, 130

Oregon, 23, 167, 174, 179-180, 182, 184, 188

orphan drug tax credit, 109-110, 114-115

Other Observation Countries ("Special 301"), 104, 106

overtime laws, 157

P

Pacific Telesis, 95

Pakistan, 50, 65

Panasonic, 93

Paraguay, 104

patents, 87, 96-98, 100, 105-107
see also submarine patents

Pennsylvania, 179, 184, 188

pension plans, 139-140

permitting (state and local permit process), 161, 167-169, 171, 189, 191, 198, 200, 202, 208, 210, 220

personal property tax exemption, 183-184
see also inventory exemptions, average property tax rates, abatements and other incentives

piracy (intellectual property), 65, 87, 97, 99-106, 108

plant site selection, 199-209
—comparison: Austin, Texas vs. Sunnyvale, California, 202-203
—domestic, 199-201
—international, 206-209
—software, 204-205

Polaroid, 156

Political Action Committee (PAC), 212

political contributions, 211-213

preferences(customs), see Generalized System of Preferences

price fixing, 88, 90, 93

Price Waterhouse, 91

Priority Foreign Country ("Special 301"), 104-105

Priority Watch List ("Special 301"), 104

Private Securities Litigation Reform Act of 1995, 72

Portugal, 55, 58-59

product liability, 69, 75-76, 86

profit sharing plans, 139-141

property tax, 30, 173, 180-184, 189-190, 196, 204
—abatements, 137-138, 146
—burden, 137-138
—depreciation schedules, 137
—personal, 183
see also personal property tax exemption

Proposition 13 (California), 162, 180

Proposition 62 (California), 163

Proposition 211 (California), 73

Puerto Rico, 47, 135

punitive damages, 75-76

Q

quality improvement teams, 149, 156

R

repetitive motion injuries, 158

research and development (R&D), 27, 29, 42, 57, 87-89, 112

research and experimentation (R&E), 111-114, 120, 127, 175

retirement plans, 139-141, 148

reverse discrimination, 150

Rhode Island, 166, 179, 184, 188

Rounds (GATT), see GATT Rounds

"runaway" plants, 131

Russia, 23, 28-29, 36, 50-51, 56, 105, 207-209

S

Sacramento, California, 202

Sacramento County, 190

Safe Drinking Water Act, 80

sales and use taxes, 185-188
—state rates, 187-188
see also manufacturing exemptions

sales tax exemptions, 185-188

same condition drawback (customs), 47

San Diego County, 190

San Francisco, California, 79, 190-191

San Francisco County, 190

San Jose Mercury News, 156

San Jose, California, 191

Santa Clara, California, 165

Santa Clara County, 190

Santa Clara County Manufacturing Group (now Silicon Valley Manufacturing Group), 163

SBC Communications, 91, 95

Section 127 (Internal Revenue Code), 139, 146-147

Section 301 (Trade Act of 1974), 33-38, 68, 87, 100, 103-106, 108

Section 337 (Tariff Act of 1930) 99-101, 108

Section 401 (Internal Revenue Code), 139-141

Section 423 (Internal Revenue Code), 141

Section 482 (Internal Revenue Code), 132

Section 936, (U.S. Possessions Income Tax Credit), 135

Securities and Exchange Act of 1934, 74

securities law, 72-75

SEMATECH, 88-89

semiconductor,
—industry, 35, 38, 41, 45, 61, 96, 103, 111, 168, 180
—Semiconductor Chip Protection Act, 87, 102-103
—Semiconductor Equipment and Materials International (SEMI), 214

Serbia, 65

sexual harassment, 149, 151-153, 160

Sherman Act, 90, 93

short supply provisions (trade), 44-45

Silicon Valley, 79, 82, 159, 162-168, 180, 190, 202, 204-205

Silicon Valley Manufacturing Group, 163

Singapore, 106, 207-209

Smoot-Hawley tariff law, 65
see also Tariff Act of 1930

soft money (political contributions), 212

software, 110, 115, 130, 155, 185, 195, 197-198, 205-206, 219
see also encryption, export controls, software industry, plant site selection, piracy

Sony, 43

South Africa, 206

South Dakota, 179, 184, 188

Southern Pacific, 91

Spain, 55, 58-59, 206

"Special 301" (intellectual property), 87, 103-106, 108

SRAMs (static random access memory), 44

State Department, see United States Government

state economic development, 198-199
—non-tax enticements, 198-199
—tax enticements, 199

state and local taxation, 173-196

state income tax rates, 174-179
—corporate, 84, 116, 175, 178-179, 209
—personal property, 182, 183, 185, 204
—sales and use, 173-175, 185-187, 189-190
see also apportionment formulas

static model approach, 117
  see also capital gains

statute of limitations (product liability), 76

statutory protection periods (intellectual property), 107

Stauffer Chemical, 23

steel industry, U.S., 45

stock options, 139, 142-144, 148
  —incentive (ISOs), 142-144
  —nonqualified (NSOs), 142-144

stock purchase plans, 139, 141-142

strategic combinations (antitrust), 91

stress claims, 158

submarine patents, 87, 96-98

Subpart F income, 130-131
  see also tax havens, controlled foreign corporations, manufacturing exception, "runaway" plants

subsidiary (foreign), 130-134

Sunnyvale, California, 165, 168, 197-198, 202-203

Super 301 (trade), 37-38, 68

Sweden, 55-56, 58

T

Taiwan, 36-37, 39, 43-44, 65, 106, 124, 126, 207-209

Tariff Act of 1930, 65, 99
  see also Section 337, Smoot-Hawley

tariffs, see customs

taxation, see federal taxation, international taxation, state and local taxation, employee compensation

tax enticements, 199

tax havens, 130

tax sparing provisions, 136

tax treaties, 136-137

Teamwork for Employee and Management (TEAM) Act, 156

technocrats, 215

telecommunications, 36, 51, 87, 94-96, 167
  —Telecommunications Act of 1996, 94

Texas, 98, 102, 151, 165, 167, 169, 174, 177, 179-180, 184-185, 188, 197-198, 202-203, 212

Texas Instruments, 44, 96, 99-100

Thailand, 28, 37, 106-107, 136, 207-209

Title VII (sexual harassment), 151

Tokyo Round, 53

Toshiba, 43

Trade Act of 1974, 34, 64, 104
  see also Section 301

trade secrets, 87, 101-102, 105, 107, 108
  —misappropriation of, 101-102

trademarks, 105-107

transaction value (customs), 46

transfer pricing
  —arms length (Section 482), 132
  —intercompany, 132-134
  —international, 132-134

Treasury Department, see United States Government

Treaty of Rome (1957), 55

treble damages, 94

tuition reimbursement plans, 146-148

Turkey, 105, 207-209

tying arrangements (antitrust), 90, 92-93

U

Uniform National Standards Act (lawsuits), 73

Union Pacific, 91

unions, see labor unions

unitary tax system, 173-177

United Kingdom, 55-56, 176, 207-209

United States Government
—Commerce Department 38, 41-42, 48-49, 52

—customs territory, 135
—Defense Department, 48, 49
—Energy Department, 48
—Environmental Protection Agency, 83
—Justice Department, 91, 93-94
—Labor Department, 156
—Supreme Court, 150, 163, 194

United States Trade Representative (USTR), 35, 37, 66-67

U.S. international trade acts, see NAFTA, Omnibus Trade and Competitiveness Act of 1988, Section 301, Section 337, "Special 301", Super 301, Tariff Act of 1930, Trade Act of 1974

U.S. Possessions Income Tax Credit, 123, 135

U.S.-Israel Free Trade Area, 65

U.S.-Japan Semiconductor Trade Agreement, 35, 38-39

Uruguay Round (trade), 36 , 45, 54

use tax, 118, 173-175, 179, 184-192, 194

Utah, 179, 184, 188

utility costs, 161, 166, 189, 198, 200, 203-204, 208

V

V-chips, 95

valuation (customs), 46

value added tax, 119-121

Vermont, 179, 184, 188

Vietnam, 65

Virginia, 179, 184, 188

Virgin Islands, U.S., 128, 135

Wal-Mart, 181

Washington State, 62, 174, 179, 184, 188

Wassenaar Arrangement (export controls), 51

Watch List ("Special 301"), 104, 106-107

Western Samoa, 135

Winbond, 44

worker safety, 121

workers' compensation, 158

workplace issues, 150-160

World Trade Organization (WTO), 34, 36, 48, 53, 56, 64, 66, 103-104 see also GATT

WorldCom, 91

wrongful termination, 149, 152-154, 160
see sexual harassment

Z

Zenith, 92

ORDER FORM

Fax orders: (650) 322-5505

☎ Telephone orders: Call Toll Free 1(800) 473-1761.
(Have your AMEX, Diners Club, JCB, Mastercard, or Visa ready.)

☞ Postal Orders:

OLIVE HILL LANE PRESS
2995 Woodside Road, Suite 400
Woodside, CA 94062

Please send copy(ies) of Beyond High Tech Survival @ $24.95 each.
School and corporate discounts available for orders of 5 or more. Please call toll free number for prices. Ship to:

Name_____

Title _____

Company _____

Address _____

_____

City _____ State: _____ Zip _____

Telephone (_____) _____

Sales tax:
Please add 8.25% tax for books shipped to California addresses.

Shipping/Handling:
Book Rate: $6.00 for the first book and $2.00 for each additional book, up to five copies. (Surface shipping may take three to four weeks)
Air Mail: $7.00 for the first book and $3.00 for each additional book, up to five copies.

Payment:
Check payable to: Olive Hill Lane Press (U.S. Funds Only)

Credit card:　❏ AMEX ❏ Diners Club
　　　　　　　❏ JCB ❏ Mastercard ❏ Visa

Card number: _____

Signature: _____ Exp. date:_____

ISBN 0-9655769-1-4